CARIBBEAN
THEOLOGY

CARIBBEAN THEOLOGY
PREPARING FOR THE CHALLENGES AHEAD

EDITED BY

HOWARD GREGORY

Canoe Press University of the West Indies

Barbados • Jamaica • Trinidad and Tobago

Canoe Press
University of the West Indies
1A Aqueduct Flats Mona
Kingston 7 Jamaica W I

ISBN 976 8125 09 8

99 98 97 96 95   5 4 3 2 1

CATALOGUING IN PUBLICATION DATA

Caribbean theology : preparing for the challenges
    ahead / Howard Gregory , ed.

    p. cm.
    Includes bibliographical references.
    ISBN 976 8125 09 8
    1. Theology – study and teaching – Caribbean Area –
    Congresses. 2. Church consultations – Caribbean Area.
    I. Gregory, Howard.
    BV4020.C66 1995   291.7   dc-20

Book design by Futura Graphics Ltd.
Set in 10.5 / 14 / Garamond / Optima
Printed and bound in Canada

This book has been printed on acid-free paper

CONTENTS

LIST OF ACRONYMS

AWOJA Association of Women's Organizations in Jamaica

CADEC Christian Action for Development of the Eastern Caribbean

CARICOM Caribbean Community

CATS Caribbean Association of Theological Schools

CCC Caribbean Conference of Churches

CPDC Caribbean Policy Development Centre

DEI Department of Ecumenical Investigations

DWME Division of World Mission and Evangelism

EEC European Economic Community

FEET Evangelical Faculty of Theology (Managua, Nicaragua)

IMF International Monetary Fund

ITLD Institute for Theological Leadership Development

NAFTA North American Free Trade Area

NGO Non-governmental Organization

PNC People's National Congress

TEE Theological Education by Extension

TEF Theological Education Fund

UTCWI United Theological College of the West Indies

UWIDITE University of the West Indies Distance Teaching Enterprise

INTRODUCTION

Howard Gregory

The notion of a consultation on theological education is not entirely new to the Caribbean. Indeed, one of the earliest manifestations of a consultation forum was staged in 1954. At a meeting in Rio Pedras, sponsored by the Theological Education Fund of the World Council of Churches, ecumenical theological education was a primary concern. At that meeting it was decided that theological education was to be seen as a vehicle for the development of the Caribbean and, further, that this goal was best achieved on a cooperative basis.

It is clear that a significant motivation for the future direction of ecumenical theological education came from that meeting. In many ways, this forum may be regarded as an imposition from outside the Caribbean as it arose out of a concern of external funding agencies about the most effective means of financing theological education in Latin America and the Caribbean.

The timing of this event was, however, quite significant. It was undertaken in a historical, social and political context in which political federation for the former British colonies was being pursued. This context, with its focus on Caribbean-wide development, made the recommendations of this meeting more palatable for Caribbean people. Significant changes in theological education followed in that same year.

Later in 1954 additional buildings were constructed on the Caenwood campus, Jamaica, to mark the creation of the United Theological Seminary. This venture involved the Presbyterians, Congregationalists, Disciples of Christ, Moravians and Methodists. The significance of this development cannot be overlooked. It marked, firstly, an attempt to strengthen the level of preparation of candidates for the ordained ministry and, secondly, the widening of this thrust in theological education to include a more representative spread of Caribbean congregations. Thirdly, it witnessed the inclusion of women in theological education in the form of Methodist deaconess candidates. Fourthly, Presbyterians from Trinidad and Tobago and Guyana, as well as Congregationalists from Guyana, were now a part of this wider Caribbean effort. Thus this ecumenical effort was taking on a truly Caribbean character.

With the breakup of the West Indies Federation in 1961 the question of what should become of Caribbean ecumenical theological education arose. A Caribbean decision was taken that the church should not be a part of the splintering of the Caribbean but should remain a uniting force.

The second event which may be classified as a consultation took place in 1961. At a meeting of the International Missionary Council held at Caenwood, "there was general agreement on the need for wider cooperation in ministerial training and the need to associate this training with the University of the West Indies". Horace Russell attributes the birth of the idea of the United Theological College of the West Indies to this meeting

As a result of this meeting, discussions were initiated with the University of the West Indies about the possibility of offering a degree in theology. It took some time for this proposed degree to materialize, but the Licentiate was ready to be offered to the ministerial students in a fairly short time. This step toward academic certification marked a significant step in the development of theological education in the Caribbean.

In the same year, the Division of World Mission and Evangelism (DWME) of the World Council of Churches carried out a survey and site visit of Latin America and the Caribbean to evaluate ministerial training in these regions. This body found that there existed a need for a united effort in theological education. In their attempt to execute their mission they received a recommendation from representatives of the University that a Theological Commission be established, made up of delegates from the three existing colleges at the time. These were Calabar, St Peter's and United Theological Seminary. The Commission was not established at this point, but rather, a Consultation which undertook the work in earnest.

In 1961 the Theological Education Fund (TEF) of New York sent a team to Jamaica to see what help they could offer in the furtherance of the decisions and recommendations of the DWME and to support the efforts of the Consultation. The TEF made a consultant available to assist the Consultation in its work. The Consultation which was born out of these developments became a much more formal and ongoing structure.

In 1961 the Consultation, which had been meeting for two years, prepared a memorandum to be sent to all the synods of member churches. The memorandum contained seven areas on which agreement had to be established if the College was to become a reality.

Having received the required endorsement, the Theological Education Consultation convened on 12 May 1964 for the purpose of constituting the United Theological Commission. Thus, while the Consultation ceased to exist, its work was continued and concretized in the establishment of the United Theological College of the West Indies (UTCWI).

The major contributions of these early consultations were in areas of increasing the scope of ecumenical theological education and in seeing theological education as an instrument for the development of the people of the region, thus defining theological education in ways beyond the narrow "spiritual" categories to which some would assign it. Academic standards also became the subject of attention.

The birth of the UTCWI marked a new phase in theological education in the Caribbean as the academic standards and opportunities were widened to embrace a greater number of students. The Licentiate and the Bachelor of Arts (Theology) soon became normative for students. In addition, students were no longer studying only classical theological subjects but were also studying social sciences and other courses offered at the University.

One problem, of course, was that the Caribbean context in which ministerial formation was taking place was also changing. The Black Power movement led to a questioning of some of the structures of authority, including the role of the minister. At the same time, ministerial candidates were also asking about the relevance of church music to Caribbean peoples. In the early 1970s there was a birth of indigenous church music in which the UTCWI students became involved. In addition, there was an interest in the relationship between the arts and religion in the Caribbean context.

Back in 1967 there was a growing awareness of the need for new expressions of ministry in the Caribbean. Accordingly, a project called the Mandeville Project was planned to involve faculty, students, a sociologist, and community representatives in an examination of the impact of bauxite mining on the community of Nain. This early attempt ended with the College retreating as the multinational bauxite company had objections to the project. It represented, however, an attempt to take theological education beyond the walls of the College and to relate it to the life of the community.

A conference sponsored by the College and the Theological Education Fund, entitled "New Forms of Ministry", was held as a follow-up to this project. This constituted a third consultation on theological education.

Among the recommendations of the conference was the need for improved facilities for training in practical fields such as:

- administration of local churches
- training of lay people at the congregational level
- various chaplaincies (schools, hospitals, prisons)
- counselling, including marriage counselling
- stewardship, including the proper management of personnel and church finances; perhaps a course in bookkeeping
- home nursing and first aid
- communication not only through speech training but also through the use of mass media: the press, radio and television
- the use of group dynamics
- musical training, and perhaps some training in the fine arts
- school management for those who may be called to render this service

This Consultation represented an attempt to wrestle with the needs of ministry in the Caribbean, but sought to do so by providing a more practical form of training for ordained ministers. Therefore, while attempting to wrestle with the "how" of ministry, it did not address in any serious way the theological and philosophical issues which the context was raising for those involved in theological education. This is not surprising however, as the leadership of the churches, the leadership of theological education, and the funding of the same were still not wholly in Caribbean hands.

For over two decades the work of the UDCWI, as well as that in other Caribbean seminaries, was guided by the kind of concerns raised in 1967. Increasingly, it has become clear that the pattern of theological education which has become dominant is inadequate for the Caribbean region. It has become difficult, however, to say exactly what the problems are and what needs to be done. In this context of uncertainty, criticisms have surfaced from the churches about the ministers the College produces while the College has verbalized its share of criticism of the churches for perceived shortcomings. It was clear that this situation of accusations and counter-accusations was not creative and did not have the potential for forward movement.

At other levels new creative reflections were taking place within the Caribbean. First, questions were being raised about theological education as something for the selected few to be ordained and the extent to which provisions ought to be made for a more encompassing approach which embraces more members of the community of faith. Secondly, given the dynamic nature of theology and the increasing recognition of the contextual nature of theology, a concern which arose was the extent to which theology as taught in the Caribbean reflects that reality. Works on Caribbean theology have been produced for some time, but the question was being asked, to what extent and how does the Caribbean attempt to expose the students to this way of doing theology?

In the wider church community there is much talk today about globalization of theology and theological education. The College, as the leading centre for theological education in the English-speaking Caribbean, needed to reflect on these matters and determine the extent to which these are issues of concern for Caribbean peoples.

For a long time one has been hearing references to the fact that the College is not adequately preparing men and women to minister in the Caribbean context. In this connection, the academic preparation is often acknowledged but there is a claim that ministerial formation is deficient. Here the focus is really on the practical aspects of the pastoral office. The UTCWI, as a new venture and an ecumenical one, is often blamed for any shortcoming in this area. The question is, is there some model or approach from yesteryear which the College could just recapture to produce adequately trained pastors, or is it that the people of the Caribbean and the needs of the Caribbean society have so changed that both College and church need to wrestle with the kind of ministerial formation process which will be adequate for the job?

During a visit to the Caribbean of the Revd Dr Keith Rae of the Division of Global Ministries of the United Methodist Church, USA, a meeting was held with the President of the UTCWI to discuss developments in theological education and the status of the College. At this meeting the idea of a consultation on theological education was aired. Dr Rae embraced the idea and expressed a willingness to assist with financial support for the staging of such an event and to lobby for support in various other fora for us.

A decision was taken to hold a Consultation on Theological Education in the Caribbean, 26-29 January 1993, on the College campus. This Consultation brought together representatives of seminaries in the Caribbean covering English and Spanish-speaking areas, heads of communions represented in the College, student representatives from the various seminaries and representatives of international partner agencies who support the work of the College. Funding came from the United Methodist Church (USA), the Methodist Church Overseas Division (London), the Council for World Mission, the Programme on Theological Education (Geneva) and the United Church of Canada.

This Consultation explored the following issues:

- The status of Caribbean theology
- The methodology of Caribbean theology
- Caribbean theology and globalization
- Caribbean theology and missiology
- Ministerial formation for the Caribbean churches

It was anticipated that material coming out of the Consultation would provide some new perspectives and directions for what we do by way of theological education and ministerial formation at the UTCWI.

The question may be asked, what is distinctive about this Consultation? In answering this we can point to a number of features. First, unlike previous gatherings of this nature, this Consultation represented the gathering of a truly Caribbean leadership of the churches and theological colleges to address issues of concern to the Caribbean church. As an institution the church was not exempted from the impact of colonization, one manifestation of which was the retention of the leadership of the churches and theological colleges in expatriate hands up to a few decades ago. Secondly, that Caribbean theology, as a distinctive mode of theological reflection and praxis, could be the basic assumption and background against which the Consultation was being held represents a tremendous forward movement of consciousness and affirmation of Caribbean peoples and the Caribbean reality. Thirdly, there was a clear recognition that change in the leadership of the churches and the theological colleges has not brought an automatic indigenous understanding and expression of ministry, but rather, the perpetuation of inherited colonial models. Fourthly, there was a clear affirmation that the development of Caribbean theology and a contextually relevant approach to theological education require a Caribbean wide ecumenical strategy, thereby negating the forces of division which have often been operative in state and church. Fifthly, there was a recognition that there is a certain urgency about the task of reflection on the issues of the agenda if the Caribbean church is to be ready to face the challenges which confront us as we approach the twenty-first century.

The layout and sequence of the chapters in this volume have been determined by the order of events at the Consultation. It is regrettable that not all the contributions are presented in this volume as it proved impossible to get some contributors to submit their edited scripts even after repeated extensions of deadline.

The first chapter contains the keynote address given at the official opening of the Consultation and seeks to define the nature of the challenges which the twenty-first century poses for Caribbean theology. There is a recognition that this is not some kind of futuristic exercise but is inherent in the present situation. It is indeed our engagement of the challenge inherent in the present that will determine the level of readiness for facing the twenty-first century. Theological education is seen as the arena and process within which the readiness is nurtured.

The first task, then, for those in theological education is to identify the main Caribbean priorities. This calls for more than just a mirroring of what is. It demands a vision of a "preferred Caribbean society" with the economic and social props necessary for the maintenance of the same. Realism with regard to the odds to be faced in such a process must be the hallmark of this exercise.

Education, for Ham, is central to this process as it is through education that this vision of a preferred Caribbean society will become a reality. The educational enterprise of which Ham speaks, however, is not just one which serves a functional role preparing persons for the workplace. Rather, it seeks to generate a critical consciousness as a theological concern and as an outcome for the learner.

Here theological education takes on special significance. Central to the task of theological education is the task of developing and teaching an authentic Caribbean theology. So that task of theological education is not to be seen as creating the process by which religious functionaries are produced, but the cradle within which the reflection/action process which generates the authentic Caribbean theology goes on. If taken seriously, the approach would mean a shift in the perceived maintenance perspective in the role of the seminary to that of being "the avant-garde of the thinking".

The structure of the book has been determined by the structure of the Consultation. A decision had to be taken as to whether or not major papers should form the focus of the book and other presentations and reflections fall within the Appendix. The decision to put these presentations in the body of the book reflects a position that readers would have a better sense of the flow and mood of the Consultation if this particular approach were adopted.

Burchell Taylor's Bible studies provide us with a hermeneutical approach to the reading of the Bible, specifically the letter to Philemon, which reflects the contextual reality and experience. That Taylor should break from the usual embarrassment which this book poses for some biblical scholars and explain it as a work of hope and possibility and liberation is significant.

The central thesis of this Bible study is that the voiceless, powerless Onesimus is a central figure whose protest and self-affirming act of running away from his master, Philemon, initiates a liberating process which brought a new liberating and liberated consciousness to the various parties. Thus, Onesimus achieved a new level of self-perception through this protest action and embrace of the Christian faith; Paul's conscience and consciousness were awakened in a way that led him to make a request on behalf of a runaway slave, which before seemed unthinkable; while Philemon is challenged to a new understanding of his relationship with his runaway slave which is born of the appropriation of the Christian gospel in a new way.

Taylor is clear that this epistle is not a personal and private one between three persons – Paul, Onesimus and Philemon. This letter, he argues, was written for the community of faith, one which probably existed in Philemon's household. The challenge to this community of faith in that age is used as a springboard for understanding the role of the Christian community in liberation of the voiceless and powerless through the ages. The challenge for the contemporary Caribbean church becomes clear. First, the church must awaken to the liberating potential of the community. Secondly, it must be self-critical in looking at its role in the liberation process, one which it is assumed will lead to repentance. Thirdly, with repentance will come the beginning of the renewal of the church so that it can fulfil its liberating purpose in the world.

Theresa Lowe-Ching in her presentation, "Methodology in Caribbean Theology", acknowledges the contribution of pioneers in Caribbean theological reflection and seeks to articulate current challenges and the way forward in terms of three elements – its major problematic, its essential features and its basic structure. The problematic is explored in terms of the impact of imperialism, colonization and contemporary neo-colonization on Caribbean peoples. A Caribbean theology must involve emancipation from the impact of these oppressive forces in the life and experience of Caribbean peoples. After identifying the major sources of Caribbean theology, she enumerates the major features of Caribbean theology to include articulation of the cause of the oppressed and marginalized; a focus on the transformation of persons and social structures; contextual integrity; interdisciplinary dialogue; biblical fidelity; and the presentation of a balanced approach and exercise of an option for orthopraxis over orthodoxy. The structure, it is argued, is one determined not by theoretical formulation but by praxis-centred reflection. While having a clear identity of its own, Caribbean theology needs to broaden and deepen the scope of its reflection to include further exploration of the significance of the "option for the poor" and a feminist perspective which will contribute a more holistic balance.

To those who would argue that the issue of globalization is not one that should occupy a place of significance on the Caribbean agenda, Winston Persaud's paper proposes that the Caribbean must respond to the process of globalization that is going on around us. A kind of inevitability seems to reside in this task. Nevertheless, responding to the globalization agenda does not seem a betrayal of the contextual concerns as "the contextual and global are intrinsically connected to each other". Responding to globalization also puts the Caribbean on a proactive stance, thereby negating the fatalistic and deterministic forces which have often characterized Caribbean experience. The Caribbean response must involve clarification of what the gospel is; must avoid equating praxis for sociopolitical liberation with inner healing, reconciliation and liberation; must involve a dialogue with the writers, artists and others who reflect the dynamics of Caribbean culture; must address the uniqueness of Christianity in a religiously pluralistic context; must take seriously the way in which Caribbean peoples organize and construct their world of spiritual reality; and must involve a search for the gospel "kernel" in the midst of the dialogue with affirmation of culture.

Persaud, even after advocating a search for the gospel "kernel", concludes that "Caribbean response, to the globalization of theological education, will be inevitably ideological". This development is seen as an outcome of a process of the pursuit of a praxis which reflects and reflects on the Caribbean reality. A Caribbean response to globalization must take seriously the impact of culture on the gospel in any community of faith, the sociohistorical realities which shape them, and the consequent plurality of praxis which emerges. A Caribbean response to globalization must involve an affirmation of these things if the integrity of the contextual reality and experience are to be preserved.

In planning this Consultation there was a concern that the Caribbean reality should receive a thorough and balanced treatment and not be the subject of possible esoteric treatment by a theologian. Consequently, it was decided to have a theologian and a sociologist reflect on the Caribbean reality. Noel Titus, the theologian, seeks to explore the reality in terms of its geography, history, religion and economics. In explaining these Titus underscores the importance of the sugar culture of the Caribbean, the threat of cultural penetration and the economic vulnerability of the region. On the basis of his analysis Titus then suggests a number of implications for theological education. Barry Chevannes, a sociologist, sees the dominant culture in the Caribbean as that shaped by the Africans and, therefore, explores the Caribbean reality from this perspective. His exploration covers two main areas, namely, religion and the family, with sexuality being the subject of special focus. In his exploration of the religious dimension, Chevannes argues that the African-Caribbean approach to the spiritual and the material is one which overcomes the dualistic definition of European religion and takes on "this worldly" rather than an "other worldly" definition. Implications arising from these are also explored. The African-Caribbean is also seen to have a distinctive approach to the nature and experience of God and places an emphasis on the integrity of body, mind and spirit as dimensions of human experience of God.

Turning to the African-Caribbean family, Chevannes explores these dimensions in terms of mating patterns, family forms and values underlying these. The pattern of family life which is seen to be dominant is one which begins as partners who "mate extra-residentially, then cohabit consensually and later legalize their union". That this pattern is an expression of the values of the African-Caribbean family is clearly outlined by him.

The theological concern for Chevannes centres around the way in which the contemporary church responds to this reality. A European-led church has failed to acknowledge the culture of the people of the region expressed in their religious expressions and family forms and has had a missionary stance in relation to the people of the region. The challenge for Chevannes is whether the contemporary Caribbean church is going to continue that pattern of negation of the culture of our people or to find itself and its identity within the cultural life and expression of the people.

In the paper "Ministry Formation for the Caribbean" an attempt is made to examine critically what now exists and to raise questions and make some proposals about the way forward in face of the challenges now facing the Caribbean church. A distinction is drawn between ministerial formation and theological education with the former being defined as a "sub-discipline" of the latter. The history of ministry formation as it relates to the United Theological College of the West Indies and the participating communions is reviewed before the challenges of the moment are enumerated. Among the challenges identified are:

- Determination of the nature of the ministry for which one is being prepared
- A recognition and affirmation of the individuality of candidates for the ministry
- A determination of the model of theological education to be pursued
- Determining the extent to which theological education is intended to serve a maintenance or a transformative function
- Deciding on the extent to which ministerial formation will continue an ecumenical basis or be returned to denominational turf

The matter of the determination of what we mean by a spirituality for the Caribbean and how this will be imparted to ministerial candidates is seen to be one of the greatest challenges facing those involved in ministerial formation. One theme which runs throughout this paper is the notion that the determination of the shape and direction of ministerial formation for the future does not reside in the hands of the seminary but must emerge out of a dialogue between the churches and the seminary. To this end, we will be brought back to Caribbean theology, as it is this which will inform that dialogue.

The final paper by Lewin Williams represents his effort at fulfilling a role which the Consultation assigned to him, namely, that of bringing together some of the issues and concerns which were raised during the Consultation and also to help us focus on the way forward.

What emerged from the Consultation is a clear need for a continuation of the process which began in that forum. The Consultation represented in some ways the tabling of the issues but the real work will have to be continued in the churches, the theological colleges, and in a dialogical process between the two. The planned Consultation for 1995 will determine the level of seriousness, intentionality and creativity with which the Caribbean churches and the theological colleges are prepared to invest in preparing ourselves for the challenge ahead.

1

CARIBBEAN THEOLOGY :
THE CHALLENGE OF
THE TWENTY-FIRST CENTURY

Adolfo Ham

> The political sovereignty of a people is impossible
> unless it rests upon an authentic cultural
> base created by its working people.
> The education of feeling must be at
> the heart of any struggle of liberation.
>
> — George Lamming

INTRODUCTION

I am very honoured and grateful for this opportunity. My comments and ideas are based on my involvement in theological education for over thirty years, and my service in the churches and to the Cuban people in socialist Cuba, in a radically secularistic situation which questions and puts to the test many of our presuppositions, methods and dogmas, and which also provides the context for a commitment to people's liberation.

My simple thesis is: in order to respond to the challenges posed by the twenty-first century, we have to be sure that we are answering adequately the challenges of the present time. For theological education to be effective in the Caribbean, it has to serve in the best possible way this Caribbean and our Caribbean people. It has to train the ministries of the church in order to satisfy the fundamental needs of the Caribbean people and the Caribbean church, promoting Caribbean ideas and responding to Caribbean priorities.

MAIN CARIBBEAN PRIORITIES

The publication of the final report of the West Indian Commission, *Time for Action,* has stimulated discussion and reflection in the Caribbean, among non-governmental organizations (NGOs), in the academic community, among the politicians, the churches, and all who are concerned about the fundamental issues facing us, our response to those issues, and the way in which our theological work is addressing them. Those issues which are unresolved will shape the challenges of the twenty-first century.

For instance, the Caribbean NGOs, reacting to the working document of the CARICOM Regional Economic Conference held in Trinidad in February 1991, said that the document entitled "Guidelines for Economic Development: Strategy for CARICOM Countries into the Twenty-first Century" failed to achieve five main tasks which are summarized as follows:

- to elaborate a vision of a preferred Caribbean society;
- to set out the complex elements necessary to provide the economic understanding to enable the realization of this vision;
- to indicate the role that regional cooperation and integration can play in achieving the desired vision;
- to identify obstacles and constraints which stand in the way of implementing the complexity of elements identified; and
- to identify strategies and actions which can be employed to reduce, if not eliminate the constraints and obstacles (Caribbean Policy Development Centre 1992, 9).

The NGOs' vision of a "preferred Caribbean society at the turn of the century" was one "capable of arousing the imagination and the effort of the different sectors of the society". The NGOs further saw such a society as, of necessity, comprising "social, cultural, political and environmental as well as purely economical dimensions" and striving for "a measure of internal consistency" (Ibid., 26).

This could well be a vision as well as a programme for Caribbean theology. However, I propose that we discuss what our main priorities are and then, how Caribbean theology and theological training in the Caribbean are coping with them.

Decolonization

The Caribbean nations won their independence beginning with Haiti through its glorious revolution in 1804, up to the last islands which became independent in the decade of the 1980s. We have to note those territories which still have a colonial status of some sort, mainly Puerto Rico. However as "independence" does not mean total sovereignty, decolonization similarly is a process not yet achieved which manifests itself in many dimensions: personally, collectively, politically, psychologically, and so on.

Identity

This process entails assuming our own identities, dialectically and simultaneously in each one of our countries, and as a region. It means growth into maturity, self-reliance and self-confidence. It means a clear historical vocation and conception.

Integration

Starting with the leaders of the Haitian revolution, passing through Martí, Betances and to the more recent Caribbean, statesmen such as Eric Williams, the struggle for independence was seen together with a vision of a united Caribbean. "Given its past history, the future of the Caribbean can only be meaningfully discussed in terms of the possibilities for the emergence of an identity for the region and its peoples. The whole history of the Caribbean so far, can be viewed as a conspiracy to block the emergence of a Caribbean identity in politics, in institutions, in economics, in culture and in value" (Williams 1970, 503).

Development

We must also struggle to gain a better quality of life for our people. In this task of development, we have to find and follow our own ways.

This is well expressed by Kathy McAfee in what she calls "elements of a holistic alternative to development". We need a development which:

- redefines growth;
- is ecologically, economically, psychologically and socially sustainable;
- allows women to play a central role; permits a spectrum of political and economic experiments;
- rescues Caribbean culture and identity; and
- empowers the region's poor majority and hence builds the basis of a more genuine democracy (McAfee 1991, 187).

Education

The field of general education is one of the crisis areas. One dramatic example is precisely what the report *Time for Action* recommends: the strengthening of the school system, without criticizing its presuppositions and establishing links between study and work ("learning to earn"). Is this our main goal, networking between the universities and the tertiary institutions for improvement in the the level of scientific and technological education? Is there nothing to question our very philosophy of education? What is its ultimate purpose? For whose benefit? This is clearly one of the priorities which relate to theological education, which shares so many of the same quandaries.

ELEMENTS FOR A CARIBBEAN THEOLOGY

- Our theology has to be contextual: a theology made in the Caribbean and *for* the Caribbean.
- A "decolonization" theology (developed by I. Hamid and K. Davis).
- A theology which upholds the different Caribbean cultures.
- A clear missiological content: the mission of the church as a "beachhead" for emancipation.
- An ecumenical theology: simple, clear for everybody to understand, empowering, "conscientizing" (Freire 1972).
- A theology for the renewal of the Caribbean church and the encouragement of experimental ecclesiastic structures.
- A theology in dialogue with the Caribbean diaspora and the other "Third World" theologies.
- Theological education, theological training and theological schools in the Caribbean.

In our analysis of theological education, we must follow the critical criteria developed by Freire:

- Are we teaching the kind of theology which can help us to understand our situation in order to change it through a process of reflection and action?
- Is our methodology the kind which stimulates dialogue?
- Can our students be the centres of the learning process?
- Is our theological education creative enough?

Our theological schools mirror our churches, and not the avant-garde of theological thinking. They should be cisterns for the renewal of the church and its thinking. Our theological training must prepare the ministries of the church to serve in the Caribbean, to help to develop and teach and authentic Caribbean theology, aiming also at the renewal of the churches and their mission in the area. To that end we should engage in:

- a critical analysis of our theological training
- a critical analysis of the structures and organization of our theological schools
- a serious examination of the type of ministers and pastors we are creating, the types of preachers, the type of Bible interpreters. Are we producing a good public relations executive for middle-class congregations? That does not mean that we do not have to prepare ministers for the Caribbean "intelligentsia"
- revision of the curricula which should include:
    - history of the Caribbean and of Caribbean culture
    - a sociopolitical, economical introduction to the Caribbean
    - a history of the church in the Caribbean
    - an introduction to "Third World" theologies
    - seminars on "popular reading of the Bible"
    - church, theology and Caribbean culture
    - popular religiosity, native churches and cults in the Caribbean
    - living faiths in the Caribbean
    - ecumenism in the Caribbean
    - feminist theology
- an adequate plan for scholarship in which the Caribbean Conference of Churches (CCC), for example, has a role
- a permanent fund for theological publications for the region

DREAMS AND VISIONS

I conclude quoting from Joel 2:28, a prophecy referred to in Acts 2:17. This passage has always impressed me very much, because of its psychological insight: you young people "see visions", while we the older "dream dreams". The Greek word *horasis,* "vision" in the New Testament, has a definite apocalyptic setting, taken from the *Septuagint,* evokes the visions of Ezekiel and Daniel and, therefore, the final struggle for liberation. The verb "to dream", *enhypniazomai,* appears only two times in the New Testament, one with a positive sense in this passage of Acts 2:17 and in Jude 8 with a negative meaning, those that are "hypnotized" by false ideologies and doctrines.

A great statesman once said: "I like the dreams of the future better than the history of the past." And we can remember also the famous words by Martin Luther King: "I have a dream . . . " echoed in the words of Guyanese poet Martin Carter:

> And so
> if you see me
> looking at your hands
> listening when you speak
> marching in your ranks
> you must know
> I do not sleep to dream

Blessed are those who still can see visions and dream dreams! Woe be upon those who want to destroy our ability to dream and see visions!

2

IN RESPONSE TO ADOLFO HAM (1)

Gerald Boodoo

ISSUES OF THE SEVENTIES ECHOED IN THE NINETIES

What Dr Ham has presented now in the nineties is not dissimilar to Idris Hamid's attempt in 1977 *(Out of the Depths)*. Indeed, the issues of decolonization, Caribbean identity and integration, as well as development and the various elements for a Caribbean theology, outlined by Dr Ham, echo the voices of the writers of Hamid's publication. Perhaps three paragraphs from the introduction to *Out of the Depths* would illustrate how Dr Ham is substantially reemphasizing the call by theologians since the seventies of the elements and direction of a Caribbean theology.

> It is our conviction that the guidelines for any new missiology, will arise out the depths of the historical and spiritual experience of the peoples of this

region. This is much more than learning lessons from history. What would be involved are such issues as *recalling the religious intuitions of our people,* subjecting these to scrutiny and articulating them through the symbols of faith and life. It would also involve a re-reading of our history, particularly Church history and locating our spiritual lineage on a different historical sphere. This would mean that we cease being appendages to European history or European Church history.

The new mission-thinking that emerges here, takes very seriously the total life of man: his living and working conditions; his social and political organizations; the issues of justice and human rights. It will engage in battle with the Western world concerning a proper definition of man. It rejects the working proposition *that to have is to be, or to consume is to be, or to produce is to be.* It will explore definitions that are more proper and germane to the divine image in man. It will affirm — we participate therefore we are, we share therefore we are, we love and serve therefore we are.

The new missiology will of necessity also create new structures of worship, organization, witness, theological training and dialogue. On dialogue, it will affirm that God has been and still is at work in other religions, and that God was at work in the religious traditions of our forefathers. This issue of identity for a people who were so de-historized and de-culturized, together with the problem of the terrible fragmentation of our communities, cry out for direction and clarity (Hamid 1977, ix-x).

In this regard, Dr Ham is quite sound methodologically, and I don't think there is one of us here who would seriously object to his methodological framework for a Caribbean theology. Indeed, both Ham and Hamid seem to emphasize two basic points in their methodological framework: the need to recall the religious intuitions of our people, and to subject these to scrutiny for the purpose of articulating them through the symbols of faith and life (this aspect is found more in Hamid than in Ham), and the cry for direction and clarity in the midst of our fragmented Caribbean communities (found more in Ham than in Hamid).

If my understanding of the similarity of these methodological frameworks of Ham and Hamid, of the nineties and the seventies, is correct, it is that there has been no progress through the eighties? Have we lost a decade somewhere? It would seem that to all intents and purposes, the eighties were in fact years of lost opportunity, and this for various reasons, some of which are:

- what was understood by development, was not development for us at all

- continuous postponement of addressing the issues that confront us
- no real substantial development of local religious expressions

THE LOST DECADE

Increased economic viability and a multiplicity of projects only served to fill the void of our Caribbean theological reflection with things and lots of busy-ness. Yet now in the nineties, all this theological busy-ness has left us no further along the road than the seventies. Dr Ham's presentation bears this out, especially in sharing the concern of Hamid as to our ideological framework of "development". Ham questions our paradigm of progress and education in terms of a "learning to earn", and Hamid speaks of a rejection of the working proposition "that to have is to be, or to consume is to be, or to produce is to be". The projects and ideas of the eighties seem to have been born out of a comfortable and expansive spirit; not one lean and hungry for a quest of identity and the urgent overcoming of fragmentation and alienation, as was the case in the seventies and still is now in the nineties.

In the continued plethora of busy schedules and diverse things that seemed to overabound in the eighties, we, our governments, our people and our churches, though taken up in projects, seem to have put the need to clarify and implement the insights of the seventies on hold (see Hamid's introduction [1977, 6-8] for his five basic points on this). Unfortunately, the situations they were meant to address did not remain suspended, but grew and became compounded. As a result, clarity has itself become a dubious commodity, and we are left repeating, though in more inclusivist terms in the nineties, the reflections of the seventies.

Despite the mandate placed upon us by the urgent call of the seventies, in terms of fostering renewed and indigenized worship, more adequate contextual ideology and structural organization, and in creating and supporting our own organic theology, we found ourselves in the eighties, at least in the broader institutional packages we inhabit, looking more to the North. One of the effects of this seems to be the boiling down of the search for identity to a search for a consolidation of our respective power bases. And this is done at the expense of our local religious and symbolic expressions that enliven and support our formal structures.

THEOLOGY AS THE EXPRESSION OF OUR "SPACE" OF CONFRONTATION

If I am correct in my assessment of the situation, the real issue for us, then, is not a repetition of a methodological framework to which we all already adhere anyway, in various fashions and to various degrees (especially since our revision for the semester system), and which has been given articulation in the seventies, but rather (and this I think is Dr Ham's real contribution) how we as theologians and educators are to more adequately express those dreams for the future, those visions that would grant us a truly organic Caribbean theology. The vision of the seventies seems to be resurrected in the nineties, and we are called upon to honestly take up the need to recall and enshrine our own religious *intuitions* and those of our people; and as theologians and educators, to grant them some deeper clarity and focus. It is this effort that I find compelling, attractive and redemptive for the future of Caribbean theology and theological education.

In general, theology today, I would say, needs to express the space of confrontation between our limitedness, our various understandings and interpretations, our conditionality, and the object or direction of these limited expressions. This is the stuff of our visions, the expression in our times of this gap between ourselves as conditioned, and the unconditioned God. Theology needs to be able not to "fill" that space (that is the mistake of the opulent eighties) but to grant it some understandable avenue for expression.

In terms of theological education, as necessary as new and relevant courses may be to our appreciation of this "space", these courses themselves offer no adequate "description". In inadequacy lies the living aspect of theology, which now primarily becomes our, and our people's search to more adequately express this space of our indwelling. It is only in our caring perplexity and constant participation in the personal and social cares of ourselves and our people, that not only the urgency of our problems, but the hope for some resolution is possible. But what are the contours for expressing this space as *Caribbean* space?

We must not underplay the need to have fora like this Consultation to keep the dynamic of concern, and the agitation for resolutions to our problems, alive. In a region still plagued by sectionalization, we need these fora to continue to voice our dreams of the future and to seek support for these visions. Yes, there are structural problems of finance, denomination, and so on. But, as this Consultation shows, a

surprisingly large number of these can be overcome by ⸻
accomplish what is necessary. We must *want* it.

The Caribbean theological search is really for a form of ⸻
that can, paradoxically enough, encapsulate our structures of livu⸻
freedom and spontaneity. Put in other words, an expression that
encapsulates our clarity and focus with our religious intuitions and
symbolic expressions. I think this is the inestimable value of Jesus'
parabolic sayings. We too must find our own parables.

The North has found its adequacy of expression in academic
theology, even though today it is in some decay and gains momentary
life by the injection of Southern liberation and inculturated theologies.
Our challenge is to find ours. To find our "space" of confrontation with
the unconditioned, with God, even beyond God. Perhaps the real
reflection is not theology as such (as *logos* about *theos*), more than it is
the attempt to express the fracture of *theos* in our Caribbean psyche; the
cracks, the openings that grant us the space of confrontation today
inherent in our Caribbean lives in the face of the unconditioned. In
effect, this is a task greater than *theos-logos,* and denominational
boundaries, and serves to ground us in our, and our people's, search
for some adequate expression to this dimension.

3

IN RESPONSE TO
ADOLFO HAM (2)

Ashley Smith

Wherever they are observing what is taking place this evening, there are at least four of fairly recent Caribbean history who were trailblazers in this enterprise which we proudly refer to as the making of Caribbean theology.

Those brothers must be filled with a deep sense of satisfaction at the vindication of their efforts.

The names of three of those are household names to most of us; the fourth, though an authentic Caribbean person, because he practised his craft in diaspora, is known only to a handful of us.

To those who have now earned their rest, we owe an unpayable debt of gratitude for the leadership they gave as, with little more than Caribbean machetes, they cleared the path in a thickly wooded area at a time when to speak about Caribbean God-talk was to expose oneself to the possibility of being condemned or marked for excommunication by even the most nationalistic within the church.

I speak of Robert Cuthbert, the organizing genius behind the creation of the epoch-making Caribbean Conference of Churches, and out of which has come much of what we now associate with regional identity.

I speak of David Mitchell, that intellectually versatile Caribbean churchman who first challenged the Caribbean church leaders to produce Caribbean Christian literature, to meet the needs of the church, of the people, and of the fledgling Caribbean nations.

I speak of Idris Hamid who, in defiance of the reactionary efforts of very influential opponents to the nation of contextualization of God-talk, literally launched many of us into the writing of theology characterized by self-affirmation, protest against universalism, and everything else that makes received theology into an instrument of colonial domination and the subversion of the self-determined development of the people of the region. In leading us at the symposia for the production of the papers contained in *The Troubling of the Waters* and *Out of the Depths,* Idris Hamid led many of us to understand the difference between *learning other people's theology* and being a party to the falsification of our own consciousness; and *doing theology* out of our own experience of reality and like the erstwhile oppressed of the Bible re-politicizing ourselves and thereby sharing in the creation of the kind of future we can proudly call our own.

Idris Hamid, an authentic Caribbean church parent, led the way in what he most appropriately describes as going *In Search of New Perspectives.*

Another Caribbean pioneer in the recording of the regions *Voices from the Margin* was Romney Mosely, a Barbadian who distinguished himself as a credible exponent of pastoral theology with a focus on the significance of uninhibited self-affirmation.

Because the enterprise of the construction of a systematized theology of the Caribbean is as yet so new and so fragile that it is vitally important that we garner and guard very jealously all the data available, so that there may be a minimum of waste and as a little assistance as possible from those who would be our detractors.

In his keynote address presentation Dr Ham has very appropriately identified the following areas of concern for those mandated to lead the people of the region in the crafting of God-talk which articulates the authentic experience of the people of the region in their struggle against the many forms of oppression and exploitation by which our history is characterized.

The concerns are for the following: decolonization, intergration, development and education.

To these I would add family and gender relationships. Needless to say, these are all of one piece. The problems to which they are all related are all part and parcel of the legacy of forced migration, extractive economies effectively reinforced by a Christian theology, used both deliberately and naively for the maintenance of iniquitous social, political and economic status quo.

In order that this Consultation might not become an exercise in semantics or just another event that brings academics together at other people's expense it is important that the focus be kept not on the churches or for that matter on the needs or fascinations of the privileged academic but rather on the *people* in the totality of their being and the contexts in which the majority of people are scandalized by *shame* and *self-doubt*. This is an indication of worldly fatalism, weak family and place of origin, loyalty and the notion of a God who, for most of the religious, is perennially too much for a system of containment and too little for the liberalization of those who are perennially within the system.

Our efforts must be focused on the creation of a Caribbean situation in which persons know themselves as full persons — affirmed by family and community of origins, together with communities of discipleship, work, play and worship. Persons with the strength to be unreserved, self-affirming, faithful and open to others, with a firm sense of the future as the eternal source of selfhood, redemption and renewal.

This task will entail the cooperation not only of churches and those already confessing Christ, but indeed the whole community of people of all political, religious and social orientations.

Needless to say, this theology, envisaged by the dreamers referred to in the presentation, if it is to be optimally liberating and facilitative of total development, must be couched in the language, not of traditional specialists, but of all the peoples. Those who are specialist exponents of this theology need both to be fully affirmative of self and community and sensitive to the people's need to be heard as they overcome the fear of sounding foolish, and the dependency reinforced by the persistence of the fear that is such a common feature of Caribbean reality.

Lest some be tempted to become fearful that we might be involved in the sidestepping of Scripture and the faith delivered to the apostle, we must make haste to point out that it is in keeping with the understanding of Scripture and the person and work of Him who is midpoint in God's story, that we seek to be faithful in reporting of what we hear out of the context in which we live, speak and relate to God in faith, worship and service.

In the words of Idris Hamid, *In Search of New Perspectives* (1973): "Our hope is in God our Liberator, who promises to set us truly free and works for the freedom in our midst and calls us to share in his liberating activity."

4

DAILY BIBLE STUDY (1):
ONESIMUS — THE VOICELESS, POWERLESS
INITIATOR OF THE LIBERATING PROCESS

Burchell Taylor

In our church life and our practice of the Christian life, the letters of
Paul, both the disputed and undisputed, hold a place of exceedingly
high prominence. Among these letters, the letter to Philemon confronts
us with something of a contradiction as far as our general attitude to it
tends to be. No other letters are better attested to as being genuinely
Pauline, yet this letter seems to be the least known and in many ways
the least regarded. Hence, it prompts one to ask, why is this letter so
badly neglected? The reasons, as could be expected, are many, with
some more significant than others.

Let us look at some of the well-known reasons that are relevant to
our own reading of and reflections on the letter. People who have
traditionally benefited from reading and studying many of the letters

associated with Paul's name somehow lack benefit from this, the shortest of them all. This is because it does not:

- boast many or indeed any great "quotable quotes"
- deal with any profound doctrinal issues
- have any list of ethical codes that are readily available for application, or
- offer any significant insights on church order and orders of ministry

All in all then, it is a letter that is seen as deficient in valuables for the devotional, liturgical, catechetical and apostolic interests of the individual and the church.

Another reason may be that many have found this letter a downright embarrassment. The occasion of the letter itself has given rise to this. Slavery, which has troubled our modern conscience greatly, seemed not to have done so for Paul, or for that matter the early church. For some persons this is a great blot on Paul's copybook. Here he seemed to have endorsed slavery by remaining silent about the institution itself when given a great opportunity to make at least a critical comment. This is in addition to the endorsement he seems to have given the institution in certain other letters that are associated with his name — though disputed in some instances.

He counsels obedience to the master, doing it with fear and trembling and doing it as to Christ (Ephesians 6:6-8; Col. 3:22). There is also the ambiguous advice appearing in I Corinthians 7:21. There is no certainty as to whether he advises taking the opportunity for freedom or remaining as is. Whatever the advice, the institution is not questioned. Of course, the use of the slave metaphor, in terms of his own ministry and the ministry of others, further indicates that he did not consider the institution objectionable.

The apostle's attitude tends to lead some to deal harshly with him on the matter. Part of their reaction is to behave as if the letter of Philemon does not even exist. Others have rushed to the defence of the apostle and think that his attitude and action are understandable. They regard his silence as sheer pragmatic realism. The time simply was not ripe for any attack on existing social structures. It would have been foolhardy for a fledgling church to take on the impossible task of challenging a system that supported the whole social order.

Slavery itself may have had a different character based on experiences of other times and of other circumstances, including much later experiences. It is felt that at the time it was a framework which offered scope for upward mobility to positions of greater responsibility. Indeed, the scope for movement was more promising than in some other areas of life. Paul might also have been under theological constraint, believing that the end of history was imminent, hence it was unnecessary to be too concerned with matters of social ethics on the level of tackling an issue such as slavery. Concern for the individual on a spiritual level was the most important thing in the given circumstances.

Is there another reading that we may have of this letter? I agree with those who are increasingly coming to see that there is another reading. I would like to offer a reading that gives an important place to Onesimus at the outset and locate a liberating process at work that is quite instructive for us in our own context. We need to face the fact that there is probably as much not said as said in the letter and that this, nevertheless, speaks as powerfully as Paul's words.

Onesimus comes to us in this letter *voiceless and powerless*. He is simply written about. Representation is made on his behalf; he is a law-breaking runaway slave. He meets the apostle Paul, and is converted to the Christian faith and Paul thought it fit to return him to his slave master and owner. Our letter represents Paul's effort to have him converted and once again accepted. Onesimus was the passive object of Paul's negotiating skill and tact which, to the eyes of some, was at its brilliant best. Onesimus's fate appeared not to have been in his own hands. And yet, when one looks at the whole process of negotiation, it seems as if things were not taken for granted to be as they had always been. Some new factors came into play. Chief among them was the status of Onesimus. New terms were being used in relation to him (10-13; 16; 21). These are not factors to be slighted. Significant transformative potential was surfacing, with important and far-reaching implications.

Voiceless, powerless Onesimus must be seen as the initiator of all this, and as a key liberating influence in the process that was taking place. He functioned to awaken a new consciousness even as his own self-perception became enriched by his embracing encounter with the liberating message of the gospel.

Paul's conscience and consciousness were awakened by the encounter and engagement with Onesimus. Indicative of the liberating impact on him is that he was now prepared to use the language of this runaway slave and issue a challenging request to Philemon concerning Onesimus. It all began with this slave who decided to run away.

Many theories have been advanced to account for Onesimus leaving his master's house. This includes the theory that he might have stolen his master's money and absconded. This perception exists because in the minds of many this is exactly what slaves do, or are expected to do. Paul in asking Philemon if he is owed anything and suggesting that he will pay it back on Onesimus's behalf, makes Onesimus a thief of his master's money.

This is a typical explanation of this request that could have had any of many other plausible explanations. Since we are in the area of giving explanation where there is no sure evidence I think there is an important explanation for Onesimus's running away that must be reckoned with. He was simply not accepting slavery as something to which he must be subjected. He did not think that he was fated to be a slave by virtue of his class or any other feature of his humanity. His act of running away was *protest action*. It was *an act of defiance and rebellion of the human spirit against oppression and indignity.* The more the case is made that slavery at that time was without the outrages and cruelties that we have come to associate with it, is all the more reason that the act of running away would be seen as significant.

Slavery and oppression cannot be sweetened to make them genuinely acceptable to the human spirit that yearns for the freedom to fulfil its potential. Oppression in any form cannot become benevolent and cannot be reformed to make it truly acceptable to the human spirit. It is the protest of victims in the face of such convictions that often keeps apologists and accommodationists from having their own way. It is often protest taken at tremendous risk and is seen as seemingly foolish in the eyes of others.

Therefore, running away is not simply an act of cowardice as the word would seem to connote in popular understanding. It is an act of enormous courage. It constitutes a risk fraught with consequences of the worst kind for the offender and for any perceived accomplices. It shows Onesimus's distinct preference for freedom in spite of the danger involved.

Although Paul's negotiating skills seemed to have been put to great use in asking Philemon to take back Onesimus, it might also be said that apart from Paul's own convictions, born of an awakened consciousness, he had to speak the about conditions of acceptance on Onesimus's own insistence. Onesimus now had his own conditions on which he would return, while seeing the need for going back with a sense of his status now grounded in the gospel.

This voiceless, powerless runaway slave was set to create a crisis of conscience and so be the catalyst of an awakened consciousness by the action he took. His meeting with Paul in the process is the providence of God. He becomes a living testimony to the fact that oppression cannot be made acceptable to the human spirit. The yearning of the human spirit for freedom remains always a challenge to conscience that can become conformable with its absence. What, then, can we tell ourselves at this point?

- Victims of oppression and injustice who seem to be voiceless and powerless are, nevertheless, often initiators and catalysts of powerful liberating influences and impulses. They often do not get credit for this in the eventual positive outworking of the process. This is the last gasp of the injustice done against them — a direct denial of their potential as participants in the liberation project.

- The Christian conscience must be sensitive to positive leads that can come from unexpected sources and in unexpected ways. These offer possibilities of commitment to the process of liberation where this is a real issue and a real need.

- There are times when legality and morality may clash. It often happens in situations of oppression and injustice. To be automatically and immediately drawn to the side of "law and order" in such situation without keen attention to the moral and ethical issues involved could easily put the Christian on the wrong side. There is no easier way to forfeit opportunities for commitment to human liberation than this. The matter of Paul wanting to send back Onesimus to Philemon is not necessarily one of wanting to uphold legality but a recognition of the liberating and transformative potential the whole experience will have for Philemon and the community that met in his house.

- When the quest for freedom from social and political oppression is met by experience of the redemptive power of the gospel, it becomes an extraordinary event of great moment. It is an action that gives liberation its integral characteristic. Onesimus's yearning for freedom from slavery met with the liberating experience of the gospel and created an event of personal liberation with possibilities of much wider implications which went beyond all ordinary expectation.

Here in this letter, the voiceless and powerless Onesimus becomes an initiator and catalyst for a new situation with great liberating potential. To hear the voice of the voiceless, to respect the yearning of the oppressed for justice and freedom is a call and a challenge to become participants in solidarity in the struggle for freedom.

5

METHOD IN CARIBBEAN THEOLOGY

Theresa Lowe-Ching

INTRODUCTION

Just about twenty years ago, the publication of *Troubling of the Waters*[1] signalled the emergence of a theology being formulated in the Caribbean context. This collection of papers, delivered at two conferences, one in Trinidad and Tobago and the other in Jamaica, and edited by Idris Hamid, reflected both depth of insight into the major problematic of the region and a wide-ranging approach to the theological task confronting Caribbean theologians. Indeed, this seminal publication was to set the theological agenda for the region up until now.

Very recently, however, Lewin Williams raised the question of the impact of this Caribbean theology on the real lives of people and the further questions of the "what", referring to its present status, the "why", referring to the "value of persistence" with it, and the "wherefore", referring to its future tasks.[2] These questions are, indeed, timely and point to the significance of what we are about these coming days, as we pause to reflect together on our Caribbean theological enterprise. Necessarily, we must look back, acknowledge the work that has been done and critically evaluate it in light of present concerns in order to project what challenges might lie ahead as we approach the twenty-first century.

In this paper, "Method in Caribbean Theology", I am proposing to do just that, with the modest objective of possibly stimulating thought towards further discussions and development. In considering the publications up until now, I will outline briefly the lineaments of the theology in terms of its major problematic, its essential features and its basic structure. An assessment of this will, I hope, point the direction towards a more adequate response and what I perceive as the major challenge before us.

THE PROBLEMATIC

"Imperialism of the spirit is the most final and fatal subjection any people could experience. This imperialism has done and is still doing its work among us. Yet it has not completely conquered. The human spirit in the quest for wholeness bounces back in myriad ways. In the Caribbean, the search of the human spirit for freedom, wholeness and authenticity has expressed itself in various ways."

Thus Idris Hamid expresses starkly and boldly the basic problematic of the Caribbean region.

"But", he continues, "the Church in the Caribbean has been gloriously oblivious of this quest. It has not discerned it, nor encouraged and enabled it. It has even failed to see the role it has played, unwittingly at times, in the subjugation of the spirit."[3]

An indictment, of course, but no surprise because the church historically came to the Caribbean region in attendance to the colonizing powers and with a missionary outlook which equated Christianity with civilizing.[4] In effect, the European brand of Christianity was imposed *in toto* and a "theology of imposition", as Robert Moore contends, marked "this second great age" of Christian expansion.

But as Hamid further pointed out, the issue was deeper than the mere fact of foreign imports. The more serious implication was "an understanding of faith, the *expressing* of it in creeds, beliefs, and particularly worship" which was unreal and unrelated to the everyday life of the people.[5] The God who acts in history and has revealed the authentically human to be indivisibly united with the divine in the person of Jesus Christ could hardly be recognized.

Concomitantly, the humanity of the conquered population was denigrated and their personal identity and self-esteem destroyed. The only word of God that could make sense under that type of oppressive situation, as Clive Abdullah indicates, is the cry of Moses, "Let my people go".[6]

It is not surprising, then, that further analysis into the impact of European colonization and, more recently, North American neo-colonalization on all aspects of Caribbean life, has revealed the attendant major problems, identified by Kortright Davis as persistent poverty, cultural alienation and dependency, to be endemic to the region.[7] Thus, understandably the primary focus of Caribbean theology has been on the experience of the majority black population whose ancestors were torn from their native land, cut off from ancestral ties and cultural roots and, nevertheless, survived the most inhuman conditions.

Ashley Smith's early interpretation of the Black Power movement as protest against the indignity imposed upon Black people[8] and Joseph Owen's analysis of the Rastafarian movement[9] would set the stage for more extensive explorations into African values and religious heritage. We note in particular Leo Erskine's *The Decolonization of Theology*[10] and Kortright Davis's *Emancipation Still Comin'*.[11] Although Davis does identify specific crises and problems attributable to other factors such as geography or history, his major point of reference and concern remains the Black experience of oppression and its dialectical possibility of true emancipation.

The Sources

The main sources of Caribbean theology are succinctly identified by Emmette Weir as, first of all, the Bible, uses of which in sermons and statements of Caribbean groups he would recommend for theological exploration; the history of Caribbean people; the writings of certain Caribbean sociologists and economists and the history of the Church in the Caribbean and statements of conciliar and ecumenical bodies in the Caribbean.[12]

Even a cursory look at the publications of Caribbean theologians to date, a list of which is again conveniently provided by Weir, will testify to the more or less consistent and effective use of the sources mentioned earlier.

The Main Features

Briefly, Caribbean theology can be seen to be characterized by the following features. It:

- is a theology which seeks to espouse the cause of the oppressed and marginalized
- is aimed at transforming persons and the structures of society
- is decidedly contextual but is open to outside influences, provided these do not compromise its integrity
- employs other disciplines, particularly history and the social sciences, to interpret reality
- affirms the priority of Biblical revelation as a theological source;
- aims at a balanced approach, recognizing strengths and weaknesses as two poles of a continuum and
- has a preference for orthopraxis over orthodoxy.[13]

THE STRUCTURE OF CARIBBEAN THEOLOGY

In a decided reaction against the theoretical formulations of a European "theology of imposition", Caribbean theology, like its Latin American counterpart, is praxis-centred, having faith praxis as both its starting point and goals, and indeed, as the very foundation of theory. Matthew Lamb, we recall, describes this theological model as a "critical praxis correlation model" whereby the Christian message is understood to be an indissoluble unity of theory and praxis, more basically mediated through praxis than through theory.[14]

The hermeneutical circle, clearly explicated by Juan Luis Segundo as an essential aspect of the Latin American liberation theological method, is consciously employed by Caribbean theologians.[15] Hence, basic broad questions regarding the quality of human existence in the region are raised; hermeneutics of suspicion is applied to the interpretation of Scripture and the theology which bolstered the colonial oppressors and was aimed at keeping the oppressed in placid submission, waiting for salvation in the life to come. A new interpretation of Scripture is then sought to affirm and validate a new way of experiencing and living the Christian message of God's liberative intentions for the people of the Caribbean region.

Thus, in many respects we have more than a theology which is only emerging. We are able to detect the clear outlines of a theology conscious of its primarily liberative task, having at hand the tools for its formulation and employing a precise method of procedure to achieve its desired goal. Of special significance is the explorations into and elaboration of the Black experience as a hermeneutical base of Caribbean theological reflections. This has resulted in the identification of specific retentions of traditional African values and beliefs, which Kortright Davis has argued persuasively, have the potential of acting as counterforces against the destructive forces of colonial legacy and promoting its transformation.[16]

CRITICAL REFLECTIONS

Why, then, does a certain tentativeness persist in the writings of Caribbean theologians with regard to theological reflections being done in the region? We note how often we read about what Caribbean theologians "should" be about. This tentativeness, I suggest, could rightfully be coming from the realization that the vistas opened up by Caribbean theology are vast indeed, and much more research needs to be done by Caribbean theologians, especially in the areas of the various other disciplines which are being used to analyse the Caribbean reality.

Not only must scientific theories being drawn upon be understood as coherent systems in themselves, but the values they affirm must be seen to be coherent with the Christian message.[17] Hence, not only must Caribbean theologians engage in more serious dialogue with social scientists and professionals in other disciplines being drawn upon, but the Scripture scholars among us must hasten to provide us with that rereading of the Bible from our side of history which, we all agree, has to be the bedrock of our theological enterprise.[18] Besides, the task of reinterpreting specific Christian symbols is still outstanding to a very great extent, notwithstanding the initial attempts which have been made in this regard.[19]

In engaging in this task, however, presuppositions need to be more clearly surfaced and influences identified and more deliberately chosen. The fact of the Caribbean being an open society and positioned to experience and be fashioned by various influences is indisputable.

All of this points to the need for more rigorous theoretical work, not in isolation from praxis, but rather by way of safeguarding the liberative thrust of faith praxis. In this regard, Juan Luis Segundo's explication of the inevitable link between faith and ideology and his argument for an informed use of ideology, understood as a system of means and ends necessary to realize the values affirmed by faith, is instructive.[20]

Also to be noted here, is the still outstanding task of clarifying the fundamental methodological assumption identified by Latin American liberation theology in particular, and appropriated by Caribbean theology, i.e., the hermeneutical privilege of option for the poor and oppressed.

A recent publication by John O'Brien entitled, *Theology and the Option for the Poor,* [21] could perhaps, provide valuable insights in this endeavour. O'Brien argues that the perspective of opting for the poor is capable of creating the conditions for a type of Gadamerian "fusion of horizons" between the immediate, albeit less differentiated consciousness of the poor and oppressed as the privileged ones of God, and the theologian's characteristically objectified grasp of tradition. In this fusion of horizons that follows on a genuinely dialectical commitment of the theologian to the poor, he holds, "something new comes to be that both allows the Gospel to disclose itself as good news to the poor and facilitate a qualitative step forward in the methodological self-appropriation of the tradition".[22]

Following this line of thought, it becomes clear that rootedness in the praxis of liberation does not necessarily exclude the demand for theology to articulate its horizon of interpretation. Thus, the dual challenge confronts Caribbean theologians to not only express their understanding of option for the poor and oppressed in theological categories, but also to engage more seriously in a spirituality of liberation involving actual participation in the struggles of the poor and oppressed, as has been suggested time and again by Caribbean theologians themselves. What this will entail, O'Brian claims, is a "vital attitude, all-embracing and synthesizing" allowing us to break with familiar mental categories.[23] The implications of this will, I hope, become clearer as we consider what, in my opinion, is the major challenge confronting Caribbean theologians today, i.e., a serious inclusion of the feminist agenda in Caribbean theological reflections.[24] A story immediately comes to mind when I begin to consider what this might entail. The story goes:

> Once upon a time, back when the world was new and all, the Dragon lived in the deep. She was mother of all, a dark force whose domination spread from shore to shore. In passion, her awesome power shaped and sculpted the land; a cliff of granite destroyed, a dazzling white beach created. Islands and lagoons, sand bars and channels emerged in response to her restless movement. But with her smile, the seas rippled in delight. Dancing sunbeams made diamonds in the waves, and gentle swells, resonating to her pleasure, caressed the shores of a thousand islands, sending warm tides surging through quiet wetlands, the swampy nursery of all living things.

Then one day, the Dragon was lured from the sea and banished to a cave. For reasons which seemed good at the time, it was decided by the powers-that-were to put an end to the restless destruction and creation of the Dragon. Something had to be done, and off to the cave she went. As things turned out, the sea is still rolled by the children of the Dragon, but the Dragon passes sunless days confined to Stygian gloom. Should despair and anger drive the Dragon to leave her gloomy abode, the way is blocked by a guard at the gate, St George by name. This fabled knight stands watch with sharpened sword and stout spear, keeping the Dragon under control. Once, it was said, the Dragon broke loose, and the violent passion, compressed in the cave, poured out across the land. For days the Dragon raged until St George and a hastily assembled band of junior knights coralled the beast.

Quite recently, however, a strange heretical thought has appeared in the land. What if the Dragon were not the terrible beast so horrendously described in song and fable? Angry for sure, but wouldn't you be if you had been locked in a cave for a millenia? Perhaps the Dragon is only lonely. What would it mean to make friends with the Dragon?[25]

The implications of this story are obvious. The almost total lack of references to the experience of women and women's contribution in our theological reflections up until now is remarkable. Also to be noted is the dearth of women theologians themselves in our region. But could not the befriending of the dragon begin to provide a distinctly creative approach to theology which could eventually contribute significantly not only to our theological enterprise as such, but more importantly to the transformation of persons and society which is the immediate goal of Caribbean liberation theology?

Ashley Smith has already made a connection between patriarchy and domination and the type of unequal distribution of resources which pertains in the resulting hierarchical structure.[26] A basically dualistic view of reality undergirds patriarchy and, hence, the entire western world which is built upon it. The feminist approach, such as that proposed by Sandra Schneiders, challenges that world view and calls for a totally different way of seeing reality, a way of experiencing the universe as a whole.

This approach, Schneiders argues in relationship to biblical interpretation, demands a critique of the entire frame of reference of the biblical tradition that goes beyond reinterpreting specific aspects of the tradition to claim the positive and significant contributions of women and to disclaim the patriarchal distortions, a critique that goes "beyond patching".[27] Thus, she understands feminism as "a comprehensive theoretical system for analysing, criticizing and evaluating ideas, social structures, procedures and practices, indeed the whole of experienced reality . . . more than a theoretical system for criticism because it involves the proposal of an alternative vision and a commitment to bringing that vision to socio-political realization". Only this approach, Schneiders claims, would be radical enough to effect the type of transformation envisaged by the Christian message.[28]

To propose such an addition of the feminist agenda to the Caribbean theological enterprise cannot be regarded as another foreign import being indiscriminately appropriated. On the recent occasion of the tenth anniversary of the Women and Development Studies Department established at the University of the West Indies (UWI), Professor Elsa Leo-Rhynie, in her inaugural lecture as professor and regional coordinator of that chair, gave ample evidence of the appropriateness and effectiveness of the feminist movement as it has developed in Jamaica, particularly in the past decade.[29] We also know of the significant accomplishments of the various women's organizations, all the more effective under the umbrella of the Association of Women's Organizations in Jamaica (AWOJA) which provides networking and collaboration on both the practical and theoretical levels.

It is not far-fetched, then, to envision the values which are specifically emphasized in the feminist perspective, i.e., integrity, inclusion, collaboration and mutuality being employed not just as antidotes to the crisis-causing problems, such as those identified by Kortright Davis, but more as fashioning a total way of regarding and approaching the Caribbean reality. Moreover, to see these values not only as retentions of the Black culture, significant though that might be in this context, but as linked with the feminist approach, which seeks to address the most fundamental relational base of human beings, could indeed broaden and deepen the challenge of transforming all persons in the society. For as Leo-Rhynie has indicated, the feminist movement in the Caribbean is moving in the direction of commitment to the interdependence of men and women and recognizes that there is no liberation of one group without the liberation of the other.[30]

CONCLUSION

I have presented a brief overview of the theological approach which has been developing in the Caribbean region in terms of its major problematic, its main sources and features and the theological model it reflects. This has shown a substantial accomplishment to date. However, there are distinct areas needing further exploration and development if Caribbean theology is to move beyond its present status and have a fuller impact on Caribbean Christian existence.

The challenge for Caribbean theology to develop a more substantial theoretical base to inform and fashion an already clearly chosen, if not fully explicated praxis-centred approach to theology is a significant one. Specifically, the notion of "option for the poor" needs greater clarification on the theoretical level and greater engagement on the level of praxis.

However, perhaps the more important challenge by far at this present time is that of incorporating a feminist approach capable of going beyond the dualistic and oppositional structure of Western patriarchal society to promote a more integral, holistic vision of all creation. Only then will Caribbean theology become truly liberative and mature enough to impact on Caribbean existence in such a way as to point the way to a society of peace and justice and fuller Caribbean integration at all levels.

NOTES

1 Idris Hamid, ed., *Troubling of the Waters* (San Fernando, Trinidad: Rahaman Printer Ltd., 1973).

2 Lewin Williams, "What, why and wherefore of Caribbean theology", *Caribbean Journal of Religious Studies* 12, no. 1 (April 1991), 29-40.

3 Hamid, 6.

4 Cf. Stephen Neil, *A History of Christian Missions* (Harmondsworth: Penguin Books, 1964), 59.

5 Hamid, 7.

6 Ibid., 15-19.

7 Kortright Davis, *Emancipation Still Comin': Explorations in Caribbean Theology* (New York: Orbis Books, 1990).

8 Ashley Smith, "The religious significance of Black Power in Caribbean churches", in *Troubling of the Waters,* edited by I. Hamid (San Fernando, 1973), 83-104.

9 Joseph Owens, "Rastafarians of Jamaica", in *Troubling of the Waters,* 165-70. See also Kortright Davis, *Mission for Change: Caribbean Development as Theological Enterprise* (Bern: Verlag Peter Lang, 1982), 107-12.

10 Leo Erskine, *Decolonizing Theology: A Caribbean Perspective* (New York: Orbis Books, 1981).

11 Kortright Davis, *Emancipation Still Comin'*.

12 J. Emmette Weir, "Towards a Caribbean liberation theology", *Caribbean Journal of Religious Studies* 12, no. 1 (April 1991), 46-48.

13 In general, these features are common to all liberation theologies with greater or lesser emphasis on one or another feature. Caribbean theology seems more concerned with maintaining a balanced approach than, for example, Latin American liberation theology's almost exclusive emphasis on economics and Black theology's focus on racism.

14 Matthew Lamb, "The theory-praxis relationship in contemporary Christian theologies", *Catholic Theological Society of America Proceedings* 31, (1976), 154.

15 Juan Luis Segundo, *The Liberation of Theology* (New York: Orbis Books, 1976), 8-9.

16 Kortright Davis, *Emancipation,* 50-87.

17 Gregory Baum, "Ecumenical theology: a new approach", *The Ecumenist* 19, no. 5 (July-Aug. 1981), 65-78; Winston D. Persaud, *The Theology of the Cross and Marx's Anthropology. A View From the Caribbean* (New York: Peter Lang, 1991), esp. chapter 5, "Christianity, Marxism and the Caribbean".

18 William Watty, *From Shore to Shore: Soundings in Caribbean Theology* (Barbados: Cedar Press, 1981); Kortright Davis, ed., *Moving into Freedom* (Barbados: Cedar Press, 1977). See also Cain Hope Felder, *Troubling Biblical Waters: Race, Class and Family* (New York: Orbis Books, 1990). Felder's publication is an interpretation of the Bible from the Black experience.

19 Kortright Davis, *Emancipation Still Comin'.*

20 Juan Luís Segundo, *Faith and Ideology* (New York: Orbis Books, 1984), 104-13. See also *The Liberation of Theology,* 101-10 by the same author.

21 John O'Brien, *Theology and the Option for the Poor* (Minnesota: The Liturgical Press, 1992).

22 Ibid., 11.

23 Ibid., 21.

24 In proposing this, I concur with Ruether in claiming that "Feminist theology makes explicit what was overlooked in male advocacy of the poor and oppressed, that liberation must start with the oppressed of the oppressed." See Rosemary Radford Ruether, *Sexism and God Talk: Toward a Feminist Theology* (Boston: Beacon Press, 1983), 32.

25 Quoted in lecture delivered by Donna Markham at LCWR Conference, Spokane, Washington, 1990.

26 Ashley Smith, *Real Roots and Potted Plants: Reflections on the Caribbean Church* (Jamaica: Eureka Press, 1984), 62f. See also Rosemary Radford Ruether, "Women's liberation and theological perspective" in *Women's Liberation and the Church,* edited by Sarah Bartly Doely (New York: Association Press, 1970). Ruether contends that women's liberation is the "most profound of all liberation movements . . . because it gets at the roots of the impluse of domination" (26).

27 Sandra Schneiders, *Beyond Patching* (New York: Paulist Press, 1991). According to Schneiders, "Patriarchy is not one example of classism but the root of all

hierarchical relationships including not only sexism but also classism, clericalism, colonialism, racism, ageism and heterosexism" (24).

28 Ibid., 16. From another perspective, Mark Taylor argues for a postmodern theology which seeks to displace the "polar or, more precisely, dyadic foundation" of the Western theological tradition. He sees the possibility of theology drawing insights from deconstruction philosophy which, calling "into question the coherence, integrity, and intelligibility of this network of oppositions", thus typifies the Western theological tradition, "to create a new opening for the religious imagination". Mark Taylor, *Erring: A Postmodern A/Theology* (Chicago: University of Chicago Press, 1989), 11.

29 Elsa Leo-Rhynie, "Women and development studies: moving from the periphery?" Lecture presented at the Women and Development Studies Tenth Anniversary Symposium, 8-10 December 1992, University of the West Indies, Mona. See also Marlene Cuthbert, ed., *The Role of Women in Caribbean Development: Report on Ecumenical Consultation* (CADEC Publication, 1971); Nesha Heniff, *Blaze a Fire: Significant Contributions of Caribbean Women* (Toronto: Sistren Vision, 1988).

30 Rosemary Radford Ruether, *Sexism and God Talk*, 20.

6

CARIBBEAN RESPONSE TO THE GLOBALIZATION OF THEOLOGICAL EDUCATION

Winston D. Persaud

INTRODUCTION

I would like to begin my presentation with a sincere word of thanks to President Gregory and the planners of this timely Consultation on theological education within the Caribbean for their kind invitation to me to be a participant. Of course, my participation allows me the welcome opportunity to renew old friendships and to establish new ones. In addition, I view participation in this Consultation as a unique chance to explore further, and in new and creative ways, the richness and diversity of my own Caribbean heritage, which is ineradicably a part of me. Here, we have a singular opportunity to dialogue with sisters and brothers on the pressing question of what faithfulness to

Jesus Christ will mean for us in the Caribbean in the twenty-first century.

To be candid, formulation of this address has been a very difficult undertaking, far more difficult than I had anticipated it would be. A primary reason for the difficulty is that over the last few years my focus on the task of the globalization of theological education has been in terms of how Wartburg Theological Seminary, one of the eight seminaries of the Evangelical Lutheran Church in America, might respond. While my own English-speaking Caribbean heritage has been shaping my inchoate response to the question of the globalization of theological education, my primary focus has not been in terms of a Caribbean response. Now that I have been asked to do the latter, I have had to do some serious rethinking of some of my basic presuppositions concerning contextualization, globalization and mission: that it is European American and European Canadian, as well as Western European Christians who have to respond to the question of the globalization of theological education; it is they who have the problem of catching up with so-called Third World Christians, who for a long time now, have been responding to that crucial question. To some degree, this stereotype is accurate, but as a summary description, it is misleading and dangerous.

As I pondered the task I was assigned, I had no delusions of grandeur that I could definitively describe, beyond geography and history, the unique, discrete, discernible Caribbean context to which I might address myself. I include myself with those who feel that a "Caribbean context" is an umbrella term for a complex reality which perpetually eludes easy definition and grasp. This conviction played a part in generating a motley group of questions, among which were: Whose battles have we been fighting, are we fighting, and will we fight? Isn't Caribbean response to the globalization necessarily pluriform and partly disharmonious? Isn't a cacophony of sounds and voices vital, lest we end up with new forms of enslavement? What does the Caribbean have to offer? What should it receive? How should it offer, and how should it receive? Which global realities are present in the Caribbean? What are unique Caribbean realities? How do we pursue interdependence, mutuality and equality and simultaneously work on rooting out debilitating dependence? Are Caribbean cultures global and catholic? What centripetal and centrifugal forces are at work in the Caribbean? Which ones outside the region are helpful, and which are harmful? Is "Caribbean" indelibly stamped on our graduates' consciousness so that wherever they go they are "Caribbean", that is they neither despise their roots nor find it difficult to draw upon them in new places, both within and without the region? Do students need to go out in order to look into the Caribbean? What should continuing education be like for our pastors, deaconesses, deacons, and other church workers in the Caribbean?

CONTEXTUAL AND GLOBAL REALITIES

Caribbean response to the globalization of theological education must creatively balance contextual and global realities. But our response should neither be an appendix to other concerns which we consider primary, nor should we treat our own response as an appendix to the responses forthcoming from other contexts. To treat our own response as an appendix would not only be a self-inflicted insult, but even worse, it would be a gross violation of our identity as people who follow the crucified and risen Jesus Christ, whom the church proclaims as universal Lord. Because of who we are, we too are called upon to share in the responsibility of responding to the rapid globalization which is all around us. The concrete "giveness" of our world, of our own Caribbean context, is that economic realities, including the unprecedented speed in the dissemination of information, inexorably link us together. Our response should be rooted in the conviction that contextual and global are intrinsically connected to each other.

Thus, both the contextualization and the globalization of theological education are inevitable. Globalization is happening and will continue to happen, whether we acknowledge it and respond to it, or not. And it will affect us. But now, we have an unusual opportunity to be proactive and not allow the process under consideration to shape us in a deterministic fashion. We know only too well that Caribbean societies were determined and shaped by economic motivations not primarily concerned with the welfare of the majority of Caribbean peoples. But that is no excuse for fatalism, be it economic, political, cultural, or religious. Fatalism kills, and it is certainly not of the gospel.

Before we proceed any further, some definitions of the globalization of theological education are necessary. Don S. Browning, in an essay entitled, "Globalization and the task of theological education in North America", suggests that there are at least four distinct meanings to the word globalization. He writes:

> For some, globalization means the Church's universal mission to evangelize the world, i.e., to take the message of the gospel to all people, all nations, all cultures, and all religious faiths. Second, there is the idea of globalization

as ecumenical cooperation between the various manifestations of the Christian Church throughout the world. This includes a growing mutuality and equality between churches in First and Third World countries. It involves a new openness to respect for the great variety of local theologies that are springing up within the Church in its various concrete situations. Third, globalization sometimes refers to the dialogue between Christianity and other religions. Finally, globalization refers to the mission of the Church to the world, not only to convert and to evangelize, but to improve and develop the lives of the millions of poor, starving, and politically disadvantaged people. This last use of the term is clearly the most popular in present-day theological education; it may also be the one most difficult to convert into a workable strategy for theological education (Browning 1986, 43f.).

Building on Browning's findings of those four distinct meanings of globalization — evangelism; ecumenical cooperation, mutuality, equality and respect among churches; interreligious dialogue and cooperation and solidarity with the poor and the oppressed in their struggle for justice — S. Mark Heim offers five modes of analysis which crosscut those four theological priorities. Heim argues that, while Browning's summary of the meaning of globalization is accurate and helpful, "it is too limited to indicate the major dynamics involved".

Heim points out that despite the genuine and tangible commitment to globalization, including consensus on its definition of a number of theological schools (as expressed in their hiring practices, curricula, library purchases, recruitment and scholarship aid programmes), many communities have discovered "that there are sharp divisions over even the initial steps to be taken" (Heim 1990, 13). Heim notes that behind these diverse and even clashing responses lie differing modes of social analysis. He says:

> All views of globalization express or imply some kind of social analysis. There are clearly many varieties of such analysis, including ones primarily oriented toward a cultures' symbols and images (*symbolic* varieties); other types of analysis directed at intellectual systems and conviction (*philosophical* varieties); still others focused on a culture's functional structures for maintaining identity and meaning or legitimating authority (*functional* varieties); yet others attending to the economic conditions for the culture's organization (*economic* varieties); and those which take up the organization of power (*psychic* varieties) (Ibid. 14).

Because of the constraints of time and space, I cannot illustrate how any or all of these five modes may be applied within Browning's four categories. What I wish to emphasize here is that applying any one of the modes of social analysis to any of Browning's four categories to determine what faithfulness to the gospel means in a concrete context, such as the Caribbean, may result in the positing of a variety, including conflicting strategies for mission, sociopolitical or ethical-moral postures or forms of advocacy, etc. In such a situation, how do we determine which strategy or posture is more faithful to the gospel? This crisis is worsened by the fact that we need to be willing to point out where and what strategies and postures violate the gospel and, of decisive significance, we need to say what the gospel is, for in so many instances what is posited as faithfulness to the gospel calls for a change in the content, not just the form, of the gospel!

I would suggest, therefore, that any Caribbean theological response to the globalization of theological education that is contextually sound must revolve around the question of defining and confessing the gospel. The very catholicity of the Church is founded on the one gospel of Jesus Christ. But there is a contemporary crisis in defining the gospel which I would summarize thus: increasingly, gospel has become a mere nominal referent which is a negation of what is considered negating, evil, or dehumanizing.

Consequently, there is a pressing necessity (it would appear) to spell out clearly what the gospel is. Where it is spelt out, that too, falls under the very negation which it intends for another set of prevailing circumstances. The inevitable danger is that our modern usage of "gospel" will make it devoid of any critical, essential substance and meaning. This essential substance and meaning is necessarily centred in the Triune God, known and revealed in Jesus Christ in the power of the Holy Spirit. It is not centred in some monarchic or egalitarian theistic referent, a mere human projection, which arises out of interpreted subjective need and which, consequently, has no fundamental, evangelical connection to Jesus Christ, God the Son who became human, suffered, died, and rose again. It may be argued that what we have today is a better and clearer grasp of what the gospel is not, rather than what it is! It is far easier to find agreement on the fact that we are to be judged by the gospel than on what the gospel is.

Where should a Caribbean response begin? We need to avoid any equation of praxis for sociopolitical liberation with inner healing, reconciliation, and liberation. Let me hasten to add that I do not wish to promote a false and unliberating dichotomy between "worldly" and spiritual well-being, that is between justice and justification. Nor do I wish to suggest that such an equation is characteristic of Caribbean theologies. Here, I simply wish to highlight the indispensable human need for inner healing, reconciliation and liberation from God.

In his essay, "An unfinished journey", Caribbean writer Shiva Naipaul poignantly describes the indispensable human need for inner liberation, even in the face of oppressive and dehumanizing poverty. Naipaul by no means argues that the need for inner liberation is more real than the need for liberation from poverty. He insists, however, that the former is more basic than the latter. Not surprisingly, he does not suggest that the inner healing to which he calls attention is what Paul means by God's justification of the ungodly or reconciliation through Christ's death. In his essay, he simply assesses his predicament and that of Tissa, a Sri Lankan. Naipaul writes:

> Was I able — as Tissa has asked — to imagine what went on inside his head?
> Only — I could have answered — by analogy with what went on inside of my own head.
> I was no stranger to the terrors he had tried to communicate. His sufferings, I divined, were not a direct result of the distempers associated with poverty, with a set of implanted ambitions strangled by straitened circumstances — bad enough though those are. Below those can lurk other, more fundamental terrors: the pervasive dread which, to a greater or lesser extent, we all share when faced with the prospect of nothingness, of formlessness, of invisibility the nightmare of dissolution Above all, we need to exist in our own eyes; we need to have some reasonably lucid idea of what we are and who we are [Tissa] had lost his way and was no longer convinced of the unreality of his existence. He was becoming invisible in his own eyes. Is there a terror greater than this? Our world is overrun with Tissas. His fate is the fate of our times. How to exist, how to become properly real — that is the question (Naipaul 1988, 128).

Theologically, our response cannot stop at liberation, if by liberation we mean freedom from socio-economic and political oppression. If we stop at the latter, how will enemies be transformed into friends? Liberation must be subsumed under reconciliation.

Moreover, we need to listen to our writers, artists, and others who mirror our cultures before us, who remind us of our wounds — of our wounded individual and collective psyches — and our creative capacities to work at healing our fragmentations, individual and social, private and public.

Caribbean response to the globalization of theological education must consider of central importance the question of the uniqueness of Christianity among the world of religions including Hinduism, Islam, Buddhism, Voodoo and other religions or quasi-religions indigenous to the Caribbean. Is Jesus Christ uniquely God incarnate and the only Saviour of all? What is our basis for positing this (or not positing it), given that many people of the Caribbean find it to be neither true nor meaningful? At the same time, those religions mentioned earlier must be considered in both their religious and "non-religious" dimensions. Consideration should not be confined to the "non-religious" dimensions.

This is one vital way in which we honour our own people and counter any fostering of contempt within the church and the society for our own people. Caribbean people are not things but human beings. We, too, are created in the image of God, and we, too, are included in God's healing and liberating embrace in Jesus Christ through the Spirit. We must renew and intensify our efforts at theologically exploring and understanding Caribbean peoples' religiosity and spirituality, our construction of our inner, spiritual worlds of meaning and survival.

The issue of contempt within the church and the society for our own people is such a crucial point that it bears illustrating. I will use a Guyanese illustration whose meaning extends beyond that specific context. The church has had a curious history of both supporting the ruling planter class and of undermining it. The independence struggle was the culmination of the liberation struggle against slavery which began more than three hundred years ago. The church has long been into a form of liberation theology praxis. The contemporary tragedy was that the liberation theology praxis was not continued beyond the dawn of nation building. The enemy had been defeated and was powerless. The only remaining enemies of liberation were those who dared to challenge the oppression of the "once oppressed".

I mention with pride that I did receive a fine undergraduate education at the University of Guyana. But there was a blatant deficit: it did not engage the contemporary political realities in Guyana. While we learnt about the failure of plantation capitalism, we did not turn our attention to indigenous "neocolonialism". For the most part, university students were a microcosm of the rest of the society, including the church, for a long time. Having uncritically spouted the liberation and anti-Communist rhetoric (some of it borrowed from our overseas solidarity partners) for a long time, the church was sadly blind to the need for vigilance against human rights abuses, as we set about the more arduous task of nation building.

The politicians ruled the day in defining for the whole society who friends and enemies were. By the time we awoke from the slumber, during which we had become willing and unwilling perpetrators and victims of oppression, we had already surrendered our power to keep alive the vision of the "not yet" of the kingdom of God. The line between allegiance to Caesar and Jesus Christ, while not unambiguous, was in many ways clear. Yet, because of religious commitment, as well as political allegiance, it took a long time for us to realize that our rejoicing (with the USA) at our salvation from the communist demon was simultaneously the undermining of our real freedom to pursue justice and dignity for all our people. In the Guyanese context, having spent our energies fighting other people's demons, we were no longer able to see our own demons, much less fight them.

By early October 1990, I had received the good news that former USA president Jimmy Carter, during his twenty-four-hour visit to Guyana, in September 1990, had extracted from the Guyana government, promises of free and fair elections in the near future, privatization of the economy, reinstitution of a free press, etc. But the way in which this good news was conveyed (and I heard it repeated in Guyana during my 1991 visit) was that President Carter had extracted from the Guyana government in twenty-four hours what the Guyanese people had been trying to get from the government for the past twenty-six years!

Our celebration has got to be at best mixed. No one can feel happy when a government's contempt for its people is so inveterate that it would hear its people's cries only through the channel of that from which we proudly claim we were ineradicably liberated. Don't get me wrong; I rejoice at what President Carter has been able to do, and I welcome the different face the USA has been getting abroad as a result of his magnificent and untiring efforts at bringing about free and fair elections in many so-called Third World countries. But I cannot hide my deep pain and sadness that our own people's efforts at bringing about just that, had been arrogantly and contemptuously ignored by our own leaders.

The church cannot avoid responding to the cultures in which it exists. But the mandate of the church is not, in the final analysis, determined by culture. The critical question is how will the church in the Caribbean engage the culture so that our dignity is enhanced not diminished? To do so in faithfulness and with credibility, the church must be vigilant that its necessary identification with its context does not so tie it to its societal structures that it is unable to give witness in the midst of change and officially masked chaos.

CULTURAL FACTORS

Returning to the question of considering the religions of the Caribbean, we all are aware that at least four other world religions, beside Christianity, are practised by peoples in the Caribbean. Due consideration must be given to the racial, ethnic, cultural and class factors which characterize these religions. Is there any basis for the thesis that one primary factor which militated against any forceful attempt at overthrowing the People's National Congress (PNC) government in Guyana over all the years of their abusive and contemptuous misrule was the strong religious fatalism that is allegedly intrinsic to forms of quasi-Christianity and folk religion, and Hinduism? Can the same be said for the forms of Christianity which the PNC government exploited to good effect in Guyana? What about Haiti and Cuba, have we given due attention to the religious dimensions of "folk spirituality"? The customary diet of non-religious explanations of Caribbean phenomena, both private and individual and public and corporate, has been too restricted for our well-being. The challenge of the twenty-first century calls for bold, creative, proactive and pluriform responses in this crucial area.

In considering the critical question of Hinduism and Islam, for example, as religions of authentic responses to authentic divine revelation, I have been pondering the texts in Luke about the centurion (7:1-10) and in Acts about Cornelius (10:1-48) and about Paul's sermon in Athens (17:16-31). What was the centurion's religion that prompted the Jewish elders to recommend him to Jesus with: "He is worthy of having you do this for him, for he loves our people, and it is he who built our synagogue for us" (4b-5). What was the religion of Cornelius, of whom it was said in Acts: "He was a devout man who feared God with all his household; he gave alms generously to the people and prayed constantly to God" (2). In uncharacteristic fashion, Paul declares that the unknown god, whom the Athenians worshipped, was the Creator and Lord of Heaven who raised "a man whom God appointed" to judge the world in righteousness (cf. 22-31). Are other religions divinely given anticipations of the full and decisive revelation in Jesus the Christ? Do these (and other) texts speak to this question?

In arguing for a Caribbean response to the globalization of theological education which would include a thorough consideration of the religions in and of the Caribbean, I am calling for the Christian community and its theological institutions to relentlessly, intentionally, and publicly pursue the ubiquitous question of whether or not there is any essential gospel "kernel" as the gospel is contextually proclaimed and incarnated in the Caribbean.

The gospel does not and cannot come without culture. But the gospel and culture are not synonymous. Is it realistic to attempt to separate the gospel "kernel" from its cultural wrappings? Is there a discrete shell, separate from the gospel, in which the gospel is to be found? Is the relationship between the gospel and culture more like an onion — when you peel it away there is nothing left? These questions are unavoidable when we consider the religious, cultural and ethnic plurality in the Caribbean. I would also include class, colour, and gender in this list. Personally, over the years I have come to realize that some of the basic questions with which I struggle theologically are rooted in the milieu — religious, cultural, social, economic, and political — in which I was born and raised in Guyana.

Any response to the globalization of theological education which claims to be Caribbean has to face squarely the question of whether or not the traits of racism and imperialism (as well as classism and sexism) are intrinsic to the gospel. Furthermore, we cannot avoid attending to the question of which imperialism is more just: the imperialism of the exclusive claims of Christian faith, or the imperialism of definitions of justice and strategies for justice? Can we really avoid all forms of imperialism? Does not faithfulness to the gospel of Jesus Christ suggest some forms of imperialism which have cultural and other implications?

Before proceeding any further, let me at least point out the unavoidable twofold question of how we have responded, and how we might respond to the interreligious and intercultural milieus in the Caribbean.

Let me say a brief word in response to the second part of this question. David Krieger persuasively argues that in the unprecedented "conflict of worlds" in the global context, we have three specific options of response: *jumping-back, jumping-over,* or *jumping-in-between* (Krieger 1991, 35). But neither *jumping-back* out of fear into the culture in which we were originally rooted, nor *jumping-over* one's culture, from which one then becomes cut off, is salutary. For Krieger, the only way forward is *jumping-in-between,* whereby different world views are brought into dialogue "by placing them into a horizon of encounter so that they can come into a fruitful and transforming contact with each other" (Krieger 1991, 37). I am still trying to critically relate Krieger's suggestions to the Caribbean, especially to the world of Guyana with which I am most familiar.

Gavin D'Costa is right when he says: "While Christians and Christian history cannot be immune to the charge of racism and imperialism, one may question whether these traits are intrinsic to the gospel" (D'Costa 1988, 221). D'Costa's conclusion to his essay on the issue of mission is most timely:

> But ultimately, if the salt is to keep its flavour, the source of these values and the source of eternal salvation must also be proclaimed. If the Christian truly wishes to share with and love the non-Christian then, as with a close friend, one's most treasured beliefs and commitment should be shared. Proclaiming the risen Christ through one's deeds, thought and words is always, and has always been, the central challenge of the gospel (D'Costa 1988, 223).

There is no liberating Christian gospel for all; it is the decisiveness and uniqueness of Jesus of Nazareth as the Christ, crucified and risen, confessed and proclaimed as God's way, truth and life for Christians only, while other revealed ways are considered equally decisive and ultimately liberating for others who faithfully follow those revealed ways. This is no idle claim, and it should be made in humility, not triumphantly. At the same time, such a paradoxically exclusive/inclusive claim is made in the face of other competing alternatives. If not this exclusive/inclusive gospel of Jesus the Christ, what will ultimately heal, save liberate, reconcile, unite and call to global solidarity?

Because it is the crucified and risen Jesus Christ whom it follows, the church cannot but be involved in the alleviation of human suffering. The crucified Lord is to be found in the hungry, the naked, the sick, the imprisoned. The individual Christian and the church as a whole bear the marks of suffering, Christ's ongoing suffering in the world. The church's role as fellow sufferer, as protester against suffering, as advocate, as individual and institution that practises mercy and charity and work for justice and peace, is rooted in its identity as people of the cross.

It cannot be otherwise. Unity of the church is derived from the one gospel and from our common following of the crucified Jesus in a suffering world. From this one gospel is born several strategies of faithfulness which do not necessarily cohere. Unity is not static. Fundamentally, it is both gift and task, the diverse strategies that emanate from the one gospel that called them forth in the first place. The gospel cannot define the common good to be pursued across religious, cultural, economic, race, gender, colour, and other boundaries. Only Christians can be appealed to, to live faithfully under the cross of the crucified.

Let me be candid. Caribbean response, like all other responses, to the globalization of theological education, will be inevitably ideological. It is hoped that it will be a theologically centred ideology, that is an ideology centred in and critiqued and judged by the cross of the crucified Jesus Christ, who continues to suffer in those who are not embraced in the stance that is made, and whose suffering is thereby made worse. How we negotiate the variety of ideological postures is a primary determinant in contributing to or countering the unity of the divided churches. One overarching, universal praxis (if it can be found) cannot guarantee unity or sustain unity. Such a praxis can only become tragically demonic as Caribbean people are trapped in insuperable bondage. Only a theological vision out of which spring several praxes, indeed conflicting ones, can truly contribute significantly to the unity of the divided churches. Of course, even such a pan-Caribbean vision is not derived from abstract speculation but from authentic engagement in and reflection on the concrete conditions of life. These are the praxes which reflect the cultures of the Caribbean in which they are shaped.

Central to any Caribbean response is a passion for the gospel, and continuing quest to faithfully define and confess the gospel of Jesus Christ. The gospel is the centre, it alone is normative. It is always culturally formulated, but it is not equal to culture. In considering the question of what is the gospel, we have to consider the question of experience as a source of truth and as a medium of truth. If not the gospel, what holds us together? We need to face squarely the issue of the domestication of God: God has become so much a part of our own house that God is unfamiliar, a stranger, to those outside of our circles.

We hear and proclaim the gospel as people living within communities that have cultural, socio-economic, political and other peculiarities. But there are only two basic truths: there is only one gospel of Jesus Christ, and there are many cultures and many socio-political and economic strategies derived from the one gospel. Indeed, it would be more accurate to speak of strategies derived from the law of love. "The gospel does not build structures but relations, and transforms them from the bottom up. The gospel is not a structure" (Hromadka 1990).

While, of necessity, there are several expressions of the gospel in a variety of cultures and contexts, there is only one gospel. Even a cursory glance at the history of the church tells the story of the ongoing struggle to know, define, and confess, in word and deed, individually and corporately, the one gospel of Jesus the Christ. The contemporary crisis in defining and confessing the gospel is a symptom of the inevitable, ongoing struggle in which the community of faith has to engage, once it persists in announcing salvation and liberation, reconciliation and renewal, in and through the life, death and resurrection of Jesus Christ.

Paradoxically, intrinsic to this "good news" is the message that God sides with the oppressed, the suffering and the "sinned against". This is good news for both the poor and the powerful. Only the God who gives life to those denied life or who experience life only in its material and spiritual distortions can give life to the powerful, the so-called non-needy and the non-poor, as well as give "life to the dead and call into existence the things that do not exist" (Rom. 4:17). It is this God who "raised from the dead Jesus our Lord, who was put to death for our trespasses and raised for our justification" (Rom. 4:24b-25). If this is not true, then the "truth" is that the powerful find their life through the creation of their own gods who safeguard their illegitimate assumption and wielding of power. By the same token, the weak, the oppressed and exploited, the suffering, indeed the whole creation, are without hope.

It is sobering to remember that praxis as strategy, as the life of faith in pursuit of holiness both emerges from and mirrors identity; it is not identity. Yet, paradoxically, since praxis as strategy mirrors identity, it cannot be separated from identity, since it participates in its substance. But neither can Christian identity be devoid of praxis. To do so would result in a docetic discipleship, that is one in which it merely appears that we are disciples of Jesus the Christ but in reality we are not.

I have already noted that there is no formulation of the gospel apart from any cultural wrappings. At the same time, cultural wrappings as form cannot be equated with the gospel. When the sociocultural wrappings are suffering and the perceived sources of such suffering, the underlying kernel of the gospel can appear to be patently divisive. How can it be that the gospel of healing and liberation gives rise to suffering that destroys rather than heals, dehumanizes rather than liberates? On the basis of the gospel, some forms of suffering are patently anathema, while other forms of suffering stem from faithfulness to the gospel. So, is there a gospel centred in the cross of Jesus Christ which transcends the "historical conditionedness"?

Suffering — including tragic forms — that is to be transcended can come from the violent imposition of forms of unity that leave little or no room for dissent, difference and individual creativity. The cross of Jesus Christ is itself the epitome of tragic and, at the same time, liberating suffering. To follow the crucified and to protest against suffering results in suffering. Hence, we distinguish between "protest" suffering and "passive" suffering as victim. It should be made clear, however, that "protest" suffering does include victimization. The crucial difference here is that the victim protests in the name of the crucified and risen Jesus Christ. To follow the crucified and risen Lord is to follow in faith; it is living out in love the justifying trust we have in him. But the lived dynamic of that trust is not a peripheral matter, merely a *diaphora*. How we live is testimony to what our freedom in Christ implies and what participating in Christ calls forth. We may make wrong judgements. Our discernment may be wrong (and more often than not it is wrong); but whether it is radical activism or quiet passivism, the path of our discipleship is inextricably bound up with our discernment of the way of Jesus.

Caribbean response to the globalization of theological education must ever be mindful that following the crucified and risen Jesus Christ is authentic to the extent that the church and the individual Christian are found among the suffering, for that is where the crucified and risen Christ is to be found. By taking on humanity, the incarnate One takes on the tragedy and reality of human suffering. He could not be human and live outside the reality of human suffering. That would be a contradiction. His solidarity in suffering is salutary and liberating. However the church and the individual Christian participate in Christ's solidarity with the suffering, solidarity with the suffering will be incomplete, indeed pretentious, if it fails to be a solidarity among the suffering in their varied "historical conditionedness". Only in this way can we speak of a genuine family resemblance among the suffering — a family resemblance that includes, but is not limited to, race, gender, skin colour, class, ideology, culture, and so on. "And the Word became flesh and lived among us, and we have seen his glory, the glory as of a father's only son, full of grace and truth" (John 1:14).

Communities of faith in Christ define the gospel contextually in the midst of sociohistorical realities which shape them and are in turn shaped by them. Unity of action, in contradiction to unity in the gospel, is ideological. But unity in the gospel does not imply a single, uniform, homogeneous response to the gospel. That, too, would be ideological, and in that case, negatively so. It is conceivable that agreement on the gospel necessarily implies plurality of praxis. If there is no place for such plurality of praxis, then the unity in the gospel is not evangelical but ephemeral and fanciful. Moreover, it is not liberating.

As it looks to the twenty-first century, the church in the Caribbean cannot avoid serious wrestling with the current phenomenal growth of Evangelical and Pentecostal Christianity in Latin America and the Caribbean. It is a phenomenon that especially, though not exclusively, touches the masses of the suffering, poor, powerless, exploited, marginalized and dislocated. It is a phenomenon of liberation, of new found "free space", new found spiritual and socio-psychic healing in the midst of fragmentation and chaos. It is a phenomenon of division, a spiritual and social separation. In short, both centripetal and centrifugal forces are at work simultaneously in the churches and the societies as a whole.

The question of the unity of the church will press itself in new and demanding ways in the near future. How shall we respond? It is salutary to remember that unity of the church is derived from and centred in the gospel. Furthermore, unity of the church calls forth a plurality of responses for the alleviation of suffering by those who follow the crucified Jesus. Unity of the church is not supra-historical. Even as eschatological reality, unity of the church implies a life of imitation of the crucified Jesus. But the very plurality of responses to discipleship, necessary though it is to the unity of the church, is a primary threat to that very unity. Is unity sustainable or realizable, given the centrifugal forces of plurality?

As a peculiar community of memory, the church, in its liturgy and worship, in the thanksgiving, recalls the drama of God's divine love affair with the world, a love affair which includes the death of the Son, as well, and his resurrection. In recalling this drama, the church is again and again reminded that the Triune God alone is the source of its life and being. Such remembering empowers the community both to see and to live in the world differently, not dominated by self-interest but freed for the pursuit of the common good which transcends the divisions and boundaries of nation, gender and class.

7

DAILY BIBLE STUDY (2): PAUL — FACILITATOR OF LIBERATION IN SOLIDARITY

Burchell Taylor

Paul's encounter with Onesimus did not benefit only Onesimus himself (who became a Christian convert and sensed that deepened and enriched freedom which was a token of his yearning for physical freedom). Paul also benefited and, as will be seen, both Philemon and the community of believers that met in his house also stood to benefit immensely with significant implications for the wider society.

The impact on Paul himself is seen to have been so significant that it is felt that had this Philemon encounter taken place earlier in his life, the results might have been seen in letters written subsequently. He might well have displayed a more openly critical attitude to the institution of slavery or at least its practice amongst Christians. The encounter seemed to have been a challenge to Paul's own conscience, resulting in an awakening or raising of his consciousness in such a manner and to such a degree that it led him to take follow-up action that extended and deepened the liberating process. It is in this light that he is regarded as a facilitator of the process and might even be rightly called an emancipator on the basis of how the matter unfolded and the ultimate direction in which it pointed.

It may seem far-reaching to make claims of this nature in regard to Paul, based on how some have seen his role in this matter and have been severely critical of it. He has been seen as compromising and accommodating, especially in terms of sending back Onesimus. However, the reading being offered here asks for a hearing. The apostle does not make the claim for himself that is now being made. There seems to be no self-conscious effort on his part to present himself in such a manner. Nor is there any effort here to whitewash the apostle and put him above critical scrutiny. It is, rather, reading the text with an eye on the impact the Onesimus encounter seemed to have had on Paul, based on the effort that the letter represents. This allows us to see certain liberating realities and possibilities emerging that cast the apostle in the light of a facilitator and even as an emancipator.

We need to notice that, as the story unfolds, the encounter and engagement with Onesimus, the runaway slave yearning for freedom, it was not simply an individual soul being saved — it was the total transformation of a human being. It was a human being assuming a new status, described in language that was unprecedented — son, the heart of the apostle, brother, one capable of being accepted as a worthy representative of the apostle himself, and one who could play the role that Philemon himself could play in the apostle's company (10, 12, 16, 21, 17, 13). Onesimus was expected to be accepted on the basis of a radically new status; a status reinforced by a play upon his name. His name means "useless" but he was now useful. He was to be seen in a brand new light (11).

This kind of approach and use of language are of emancipatory significance. For one who was diminished and demeaned by being cast in the role of slave (whatever the opportunity and privileges he might have had as a slave), to be seen, referred to and accepted in such new ways, meant liberating influences were at work and were being encouraged.

Surely one of the ways in which people have been oppressed and continue to be oppressed is manifested in the language that has been used about them and in relation to them. The language communicates a perception and, if used by the dominant in reference to the dependent, by the oppressor in reference to the oppressed, it has a way of affecting the self-perception of the dependent and oppressed. They tend to internalize the portrayal conveyed by the language and conform accordingly.

Language can be both oppressive and liberating. Those whose tool of work is primarily words must always bear this in mind. The use of words for ultimate effect is something that belongs to the church's ministry. The oppressive or liberating possibilities and effects must not be underrated. We do tell a lot about how we think of people by what we say about them, how we address them and how we describe them. The language of relational significance is especially important. We do tell a lot in terms of how we genuinely present them to others. We do communicate a lot when our way of speaking is backed up by our way of treating those of whom we speak. Paul's designations of this man say something absolutely significant. He was being regarded in a new way.

Onesimus the oppressed has been affirmed by the apostle in the way he speaks about him. Yet the apostle was not limiting this new attitude only to himself in relation to Onesimus. The process must extend to his oppressors — his former owner and the church community that met in his house.

Several theories have been put forward to account for Paul sending back Onesimus. These will not detain us here. However, one simply wishes to say that there is no basis for ruling out whatever might have been the other reasons. It might have been that Paul came under serious moral constraints — almost, if not totally, irresistible constraints — to do what he sought to do. This was not simply out of fear either for his reputation based on his association with a runaway slave or for his friendship with Philemon lest it be observed that he was harbouring his runaway slave. One senses that there was an imperative inherent in the event which resulted in his raised consciousness about the whole matter. Something ought to be done about the status of someone like Onesimus, at least within the community of faith, if nowhere else. He cannot simply go on being seen as he has always been seen. The place to begin is where he was best known. The impact on Philomen and the church community would have been significant.

Paul himself had come to terms with the change in Onesimus's status. Where better that others should come to terms with it as soon as possible than where the man was held as slave? One believes that Onesimus himself insisted that if going back at all is involved, it must be on the basis of the new sense of dignity that had come to him. The seeds of further liberation are invariably inherent in any single act and experience of liberation. It becomes a moral power to be contended with in any given situation. Those who once held Onesimus as a slave, with all his attendant disadvantages and with all the privileges it represented to them with an easy conscience, must now know the sense of freedom that comes with relating to him properly as a human being of equal dignity — one like themselves. Their own bondage must come under attack. Paul and Onesimus must become partners in this new liberating exercise which involved real personal risks for them both. Solidarity was extended to others who were in need of liberation without necassarily knowing it. They were to confront the new situation presented by an Onesimus who returned liberated and they were challenged to accept him on this new basis.

It is invariably far from easy for those who have benefited from injustice and who have become settled with the situation as it is, to face any challenge or threat to that situation. They would have built up a stock of reasons and arguments legitimizing the given situation or simply accepting that it was given, and without any needed or necessary alternative. This is why it often is the oppressed who themselves become the true teachers of freedom through their unceasing longing for freedom, their resistance, their protest and the solidarity shared with others who have caught the vision. This is why when the liberated oppressed become oppressors, it is truly an aberration. It casts doubt on the authenticity of the liberation originally claimed.

Paul, as exemplar and leader through solidarity, and Onesimus, an initiator by his original protest, each played his role in the process of liberation. We have our own roles to play and our own risks to take in the cause of liberation in the face of injustice and oppression. This is something we have to realize as faithful Christians. We sometimes become distracted and do not see that there are those times and instances when it comes down to us, as a matter of personal decision making, as to the level of participation we shall embrace in the given liberation project. There are some inescapable challenges that go with this. We must be willing to:

- let our old traditional, familiar, settled convictions and ways of seeing things and persons be challenged by new perspectives, and prompted by undeniable awareness and impact of human need for justice and freedom;
- draw upon our available resources of faith to meet the challenge that comes from response to such needs in solidarity and service;
- put our own new and changed perspectives in the service of

others. There are those who need to be challenged about their situation and the ways in which they have seen things and have been accepting the benefits of an unjust order and of other people's disadvantage, and

• face the risk involved in confronting the others with whom we might even share special relationships and interests. This must be seen as a further expression of the friendship and interest shared even though it might become threatening in another sense, since it is asking for fundamental and even costly change in the name of love and justice.

Such are some of the insights that emerge from Paul's acting in solidarity with Onesimus to further the cause not only of Onesimus's liberation, but also of the liberation of those who once held him in bondage.

8

OUR CARIBBEAN REALITY (1)

Noel Titus

INTRODUCTION

In seeking to portray what may be understood as Caribbean reality, one must, of necessity, explain how one understands the term. The term "reality" can be used to apply to whatever is regarded as having an existence in fact and not merely in appearance. This meaning will relate to the issue of sovereignty, to which reference will be made later. At another level it can be used to refer to that which underlies and is the truth of appearances or phenomena.

This is the basic understanding of this paper, which seeks to explain that which exists in this region. Caribbean reality relates to the state of existence in the region and, of necessity, has many facets, involving geography, history, religion, education, and so on, which make up the culture of our many people. This reality is at once the context of theological education and the challenge to theological educators. There is, however, a massive area which this paper cannot attempt to deal with; it is how Caribbean people understand their reality. This is an area on which writing has to be done, hopefully to enable more adequate reflection in the future.

HISTORICAL, GEOGRAPHICAL AND CULTURAL FACTORS

The region which we know as the Caribbean covers a large area and involves two groups of territories. On the one hand, there are several continental countries of the Central and South American littoral of this sea. On the other hand there are large numbers of islands, forming what Franklin Knight describes as an inclined backbone from Florida to the northern coastline of South America. These islands vary considerably in size and topography. From a geographical point of view, the Caribbean suggests separation — scattering rather than gathering.

The predominantly insular character of the region is in part reflected in an insularity of outlook among the peoples of the region. The fact that the sea separates one island from the other, and that the continental territories are on the other side of the sea from the islands, tends to accentuate the sense of separation. Strangely, therefore, the Caribbean Sea from which we get our identity is simultaneously a factor of physical separation. To interact with each other necessitates travel, and our means of travel between the islands are limited and costly. Other means of communication — by telephone or satellite — are also quite costly. There is often separation and distinction at another level, to the extent that an individual from one territory can be made to feel as much a foreigner elsewhere within the region as outside it.

It is not only geography that serves to separate our peoples. Differences in our history form both a separating and unifying influence. Historical differences form a separating influence, in that many territories have emerged from separate and competing European powers. These included — at different times and for longer or shorter periods — Spain, France, Holland, England and Denmark. Historical differences form a unifying influence in a rather unique way.

The only constant for a major portion of the population of this region is that their ancestors were forced migrants whose lot it was to endure a life of enslavement. The result is that this group was forced to adopt a new identity and new mores — in essence, to become new persons. In common with others of the region, albeit under different circumstances, our peoples are Caribbean, or West Indian, or Antillean, and not merely European, African or East Indian *per se*. One may observe that almost the entire population is an immigrant population. In a sense, it is only here that we belong.

Our history has done something else to us. Because of varying backgrounds and European associations, we have tended to be pulled outward towards others in our loyalties rather than inward towards each other. The countervailing attitude, which seeks to encourage strength in togetherness, is far from being firmly rooted in our consciousness. We see at present European nations, which had fought each other for centuries with the sugar islands as prizes, resolving to unite for their mutual self-preservation while the Caribbean territories seem unable to come together for their mutual survival.

The historical circumstances of the region are also reflected in the linguistic pluralism which is evident today. As the report of the West Indian Commission noted, the very diversity — involving Spanish, French, Dutch, and English — reflects the imperialist rivalry of previous centuries. To these major languages must be added Papiamento and French Creole. All of these in turn contribute to the rich cultural heritage of the region through literature, music, art, and so on.

One of the greatest influences on the Caribbean has been the persistence of sugar as the major agricultural product. The ownership and management of sugar estates still remain in the hands of a few persons. To date, none of this activity has become the serious concern of agricultural cooperatives or of governments. The result is that the plantation society of the Caribbean is not a thing of the past, but a present reality. If the slave element is missing, the hard-working ex-slave is still there to ensure a certain production level. Planters and merchants still play a dominant role in Caribbean economic life and, ultimately in the social relations within each society.

The persistence of the sugar culture — both as a commodity and as a means of influencing social organization — means that the reality for many of our people is one of recycling past history. The same routines, the same work relationships, the same lack of control over their products, form a familiar and frustrating pattern. This may well bear unpalatable fruit in other areas of social life. It is a fact that in the past some slaves copied various social activities of the dominant race, while others strove to retain their inherited indigenous culture, so the people of our region are divided between their own culture and the culture of others which is made more accessible by the affluence and technological superiority of extra-regional countries.

Cultural penetration, with a foreign value system, is therefore understandable. This raises a number of questions which relate to what is now being praised as our great foreign exchange earner — tourism. Are tourists courted so as to expose them to what this region and its people have to offer? Or are the people of the region being subordinated and made subservient to the interests of the tourists? To what extent might tourism contribute to the attitudes inherent in the plantation society? These questions have an important bearing on our reality as we move into the next century.

Religion

One of the most significant elements in Caribbean culture is religion. Most people in the region are devotees of one religious manifestation or another. Some of these devotees are non-Christian, including Jews, Hindus, Muslims, Bahais, and Rastafarians. Others are Christians, including Roman Catholics, Anglicans, Methodists, Moravians — indeed, a wide variety of confessions along with more recent Evangelicals and Pentecostals.

Over many centuries the Caribbean has benefited from the contributions of religious groups in terms of education, social services, and spiritual development. But the region has also suffered from the divisiveness which has been the concomitant of unhealthy denominational rivalries. Each group brought with it cultural traits and prejudices representing the national backgrounds of the missionaries. These rivalries have been exacerbated in recent times, by the entrance of certain North American groups which have sought to arrogate to themselves the name of "Christian". As a result of all this, the juvenile ecumenical movement may well be threatened.

Admittedly, this movement has made great strides from the time, even in the 1960s, when Christians of one denomination did not enter the place of worship of other denominations. There have also been advances in the area of cooperative social action. But this cooperation has not yet reached the stage of intercommunion, nor is it at all clear whether the churches consider this goal desirable. Interreligious dialogue and cooperation, because of the composition of various territorial populations, do not constitute so prominent an issue. The reality, therefore, is one of hesitancy on all fronts.

Education

Education is another factor in our Caribbean reality, one which has been under constant scrutiny in the past twenty years. In an address on the Mona campus of the University of the West Indies (UWI) in 1978, William Demas lamented the fact that our entire education system was an imitation of the British system. A major defect of our system, as he saw it, was "the lack of positive programmes for inculcating into the pupils pride in themselves and in the West Indies". While some effort has been made to address the issue, it is debatable whether changes have been made to the same extent that he envisaged. It is also debatable whether or not sufficient attention has been paid to the issue of values in education. And this is surely not to be limited to moral values, important as these are, which are assuming considerable prominence in educational debate in some territories. It raises questions about the emphasis on science and technology, with a consequent de-emphasis on the humanities.

Consideration of the issue of values raises the question of the provision of moral and religious education. In the English-speaking Caribbean, the churches are the targets of negative criticism for failing to provide moral leadership in the various communities. Two comments might be offered for consideration on this. The first is that the critics — mainly leaders in the political sphere — are themselves largely to blame. Their own disparagement of the churches, because they fear the scrutiny of church leaders, offers no encouragement to those youth who should follow the example of such persons in respecting religious leaders. The second comment has to do with the discrimination consistently practised against religious persons.

At our own University of the West Indies, at the time of writing, study of any discipline is possible at the expense of taxpayers. Taxpayers include both religious and non-religious persons, yet if a religious person is desirous of doing religious studies or theology, that person must provide the funds for such studies out of his or her own pocket. That this should be the case, while governments claim to be providing education for all its citizens, seems very strange. That Christian and non-Christian religious bodies continue their tacit acquiescence in this state of affairs is even more strange. And yet, that too, is a reality of our time.

There has been a tremendous upsurge in the demand for, and provision of, tertiary education in the region. This area is not limited to university education, but embraces a wide range of offerings through community colleges, theological colleges, the School of Continuing Studies of the University of the West Indies and other agencies. Such courses not only provide for the widening of general knowledge by individuals, but for the upgrading of skills to meet the increasing demands of our communities of work. Not only does this development require changes in attitudes to learning, it might also require changes in the approaches employed in the delivery of education. Already institutions in the region have begun the process of rethinking traditional approaches to education, and one must look forward to the changes consequent on such rethinking.

Economics

Perhaps the most stark reality of the Caribbean is our economic vulnerability. Small can be and is beautiful. But with limited resources at our disposal we are liable to become the easy prey of those whose markets we seek to penetrate. These Gullivers of the economic world are well aware that each Lilliputian lacks the capacity to challenge, far less conquer them. This was the point being made by William Demas when he argued the distinction between formal and effective sovereignty. Caribbean countries, he stated, needed to surrender some of their formal sovereignty in order to have effective sovereignty. Furthermore, it was only from the position of strength gained by such adjustment that they could hope to deal with the international economic system.

In the face of our own foot-dragging over integration, these economic Gullivers are growing in size and strength. The European Economic Community (EEC) and the North American Free Trade Area (NAFTA) are realities of enormous magnitude for the Caribbean. Even when we have negotiated with them, we cannot forget the International Monetary Fund (IMF) and the World Bank which hover in close proximity to our shores. The controlling influence which they have exercised over the region in recent years has serious implications for us. Health, education, and welfare frequently suffer under IMF and related programmes. Gains recently attained in living conditions are quickly lost because of the inability of governments to maintain essential services. In some instances, the deprivation of those living under such programmes leads to social unrest.

What challenges might this reality have for theological education in the region? This is a question which will need to be answered by all those engaged in this enterprise. What I shall try to do here is to suggest a number of concerns, which come readily to mind, in the hope that they might stimulate discussion. First of all, given the economic circumstances of the Caribbean and the lack of adequate financial resources in the churches, it would seem that greater emphasis should be placed on Theological Education by Extension

(TEE) than is currently being done. This is, in part, to recognize that teritiary education is not limited to university or seminary education, and that it does not necessarily have to include a degree or diploma as its ultimate objective. In part, it might also call for a proper examination of the purpose of theology. The purpose of that discipline, I believe, is not simply to reason in abstraction on issues considered theological. Among other things, it will involve an attempt to interpret one's reality in such a way as to suggest change, as well as provide understanding. And this will follow from in-depth theological analyses of those systems and attitudes which shape Caribbean reality.

Again, with due regard to our economic circumstances, I suggest that we need to pool our resources to make theological education more readily available to the region. It is true that the theological colleges cooperate extensively in the administration of theological programmes. But we do so as independent, separate institutions, which do not necessarily draw on each other's resources. To that extent we are largely engaged in duplicating each other's work. The cost factor seems to suggest that we share the load. For example, the demands for extension courses might be more effectively met if we operated a single programme with individual institutions covering different parts of the region. By this kind of approach, the needs of each denomination could largely be met without each having to find the resources for a whole programme. Who knows, we may well find a way of applying this to our usual tasks. This, of course, would entail a conscious effort to reduce the narrowness of denominationalism. I should not be understood to be saying that this is easy, but that we must begin to explore ways of sharing our resources on a regular basis. The adverse economic circumstances of the region should prompt an evaluation of our approaches for the future.

Another challenge which our reality poses for theological education is that of making our theology truly indigenous. Caribbean theology cannot be European theology, not even Third World theology. It needs to be a theology which responds to Caribbean reality. Such a theology should draw on Caribbean lore — its mythology, folk wisdom or proverbs — in order to explicate the experiences of Caribbean people in a language that can evoke deep responses from such people. This would mean a greater relationship between our theology and literature to enable us to discern how Caribbean people interpret their reality. And it is an aspect of our work which requires conscious and consistent effort. For such a task to succeed, an integral part of the process must be the formulation of a publication programme. This does not only involve enabling persons with material to get it published, it includes providing suitable means of publication to meet our own needs. I think it is true to say that international publishers were not originally keen to produce African and Asian theological material. Yet theological groups in these regions have successfully mounted small operations of their own, thus disseminating their indigenous theology.

CONCLUSION

Such conscious and concerted approaches to theological education might help to break down insular prejudices, and thus assist in creating a congenial atmosphere for the integration process. Our graduates should not leave with the same insularity with which they might have entered our colleges. Nor should they become each other's rivals when they have to work in territories other than those of their birth. In other words, we might have a greater role to play, *vis-à-vis* the integration process, than we might consider likely. Not unrelated to this might be the need for theological education to be further broadened to reflect our recognition of the religious plurality of the region. Undue denominational separation will increasingly run counter to the demands for integration which are becoming more strident with each passing year. We can and should learn from each other; we can and should work more closely together.

9

OUR CARIBBEAN REALITY (2)

Barry Chevannes

INTRODUCTION

In this paper I focus on those Caribbean cultures shaped predominantly by the Africans. The justification for this approach is that in terms of mass movements they were the first, and now constitute the majority. In one aspect, namely population, Trinidad and Tobago and Guyana are currently the exceptions, boasting East Indian majorities. But even here, the folk languages spoken by all the people is that formed by the Africans.

It is important to recognize this, because of the popular theory that the Caribbean is a melting pot: a little bit of Arawak, a little bit of Chinese, Lebanese and Portuguese, a large amount of India and the most of Europe and Africa. This is a misunderstanding. Caribbean reality is shaped by Africa.

For the purpose of this discussion of Caribbean reality, however, it really matters little whether we view the question of the origins of Caribbean culture this way or that. The main point is to outline what it consists of, to understand it, in order to shape our attitude to it. This I will do in two areas: religion and family, with particular emphasis on sexuality.

RELIGION

The question is not whether Revivalism can be found in Africa or not, or even the matter of spirit possession, the drumming and the dancing. It's not whether Voodoo is traceable to Dahomey, as it is, or Santeria to the Yoruba and Kumina to the Congo. That's not the point. The point is to understand the outlook of the people on the world and how that outlook has shaped their institutions.

Let me begin with the conception of and approach to the world. The best way to understand how African-Caribbeans generally perceive the world is to contrast it to the conception conveyed by European religion. In the latter, there is a material, visible world and then there is a spiritual world, preparation for which is made in this life and access is through death. This sort of dualism pervades Biblical and subsequent European theological thought. In African-Caribbean religion, on the other hand, these two worlds are really seen as one. The same world in which we live and breathe is the same world inhabited by God, the Spirits and the Ancestors. This belief produces an approach towards the world which we might call "this-worldly", in contrast to the "other-worldly" approach propounded by European religion.

It shapes an attitude to life which places a premium on fulfilment and achievement in the here and now, rather than postponement to an "afterlife". Understanding this view of religion helps us to better appreciate past strategies adopted in the forging of African-Caribbean culture, based on the will to survive at all costs, as well as to understand the present motivation to succeed which drives our people, whether athletes, politicians, entrepreneurs or mothers sacrificing to send their children to school. For if there is only one world, then only in it can the fruits of one's life be seen and enjoyed.

What this means is not, as some might wish to conclude, an inability to postpone gratification, but rather that secular values, such as upward social mobility, or the accumulation and show of wealth, are sanctioned by religious values.

It shapes, their attitude to justice. For example, African-Caribbean people are not much at home with believing in other-worldly justice as compensation for this-worldly injustice. They believe that the just character of God ensures that no evil goes unrequited, no good unrewarded, in this life.

It shapes, finally, an attitude to work which sees little merit in work without reward. When Caribbean countries succeed in transforming their economies the way the South-East Asian tigers have theirs, it will not be because we have imitated the work ethic of the latter.

Closely linked to this is the need to reveal one's social achievement and status, through symbolic statements and symbolic interaction. Consider the need to show wealth. To be rich but not to have the status of being rich is to be considered mean. To be mean is to be lacking in social grace, to be antisocial. "And what's the point", people will say, "you cannot take it with you."

Turning to the conception and experience of God, it is my understanding that African-Caribbean peoples place a greater emphasis on the experience of God as a normal part of human life than they place on dogma. This makes their religions versatile and open.

God, the Supreme Being, is the creator of the universe and the One who sustains all life and governs the world. The folk know him as "Maasa God", or the "Father". He is so immense and incomprehensible that while respected and reverenced (for example, we still swear by him — "God strike me dead, if is lie a tellin'!"). And incidentally, though he speaks through the elemental forces of nature — thunder, lightening, storms and earthquake — he is, nevertheless, not open to being experienced.

Rather, it is the spirits, or forces, themselves subordinate to God, the ultimate source of all, who interact with man. Man is himself a spiritual force, especially to the extent that he is able to be a vehicle for the spirits. This concept is expressed either in possession by the spirits, including the Holy Spirit of God or, as among Rasta, in the belief that in a profound way God is man and man God. Here is the central feeling about all religions. For if one cannot feel the spirit, even if one may not be possessed by it, there is little point to the ritual.

Thus, African-Caribbean peoples place a great value on the integrity of body, mind and spirit. The experience of God, they maintain, cannot be limited to the mind, but must also move body and spirit. Many observers, past and present, note the emotional character of African-Caribbean religious worship, but fail to grasp its philosophical foundation.

These attitudes and values are so deeply held that they also influence the behaviour of those religions derived from Europe and America. As noted by Kamau Brathwaite, the earliest missionaries quickly learned that if their sermons did not, through the various artistic devices of speech delivery, speak to the whole person rather than just the mind, they soon lost their congregations. I further opine that this is one of the reasons responsible for the rapid growth of the Pentecostalist-type religions throughout the Caribbean, especially in Jamaica, and the attendant decline of the European ones, they leave people morally sound in spirit and feeling good in body.

Incidentally, this conception of the integrity of mind, body and soul also generates concepts of sickness and ill health and well-being, that are much closer to psychotherapy than to medicine, and traditions of preventive and curative care that incorporate practices ranging from herbal and dietary care to ritual.

FAMILY

In an article published nearly forty years ago, in which he reviewed the data put out on the 1943 Census of Jamaica, Professor George Roberts (1955, 199) made the observation that, "illegitimacy rates between 50 per cent and 70 per cent signify the existence of family forms *sui generis,* distinct from the forms characteristic of European societies".

It is an irritating fact that official society, in which the church must be included, remains still blinded by European prejudice, failing to come to terms with African-Caribbean social reality, in this respect its family forms. The failure is no more than an irritant, however, because the mating patterns and social and sexual values that produce these forms not only perdure unmolested by the sermons and moral joustings, but are penetrating into the circles of the very guardians of morality, the middle classes.

It is, therefore, no more than an irritant that Vivian Panton's book, *The Church and Common-Law Union,* should be met with such hostility by church circles, though the effect of it in denying honest and devout Christians a sacramental life is very sad indeed. And what is so ironic is the author's discovery that the same churches used to recognize common-law unions, until the state in the 1840s sanctioned the recognition only of legal marriage. Now that the Jamaican state has, in effect, given sanction to common-law, first, by the removal of the status of illegitimacy (several Caribbean countries still retain the legal restriction) and, second, by establishing the right of a common-law spouse of at least five years to inherit, it remains a double irony that the church has failed to follow suit.

Mating Pattern

I would like to outline briefly, the mating pattern and family forms, and some of the values underlying them. That a mating pattern exists is only just becoming widely accepted by people who have found it anomic and pathological. The anomie and pathology disappear when we take a diachronic look at people's mating behaviour. People first mate extra-residentially, then cohabit consensually and later legalize their union. Thus, we have the three types of mating unions: visiting, common-law marriage and legal marriage. The incidence of visiting is highest during the early years of adult life, say between 18 and 22, declines rapidly thereafter and almost vanishes after 50. Common-law, from a very low incidence in the earliest years, peaks quickly in the mid to late 20s and declines steadily thereafter. Marriage, virtually non-existent in the early years, rises steadily but not sharply to become the single largest type of union by mid 30s and early 40s and steadily to increase with increasing age.

What annoys the mind for whom Europe is the model and ideal, is the fact that the visiting and common-law unions are broken at will, leaving women without husbands; children without fathers; children, women and children with grandmothers; children with stepmothers; women with multiple baby-fathers and fathers with multiple sets of children. Instability, is what such a person sees.

What is overlooked is the movement towards stability. By the time they get married, most couples would have been visiting or cohabiting for a long period; they seldom if ever divorce. From the unstable, people progress to the stable.

What is also overlooked is the fact that only a minority of families are female-headed, the overwhelming majority being male-headed; or that even in the latter, it is the mother on whom the affective life of the family is centred. Overlooked also is the fact that most children are raised by fathers and stepfathers. In other words, what is overlooked is the fact that there is a system.

Underlying Values

It is vitally important to understand some of the underlying principles at work here. First, let us look at sexual behaviour, since, here in the African-Caribbean, it is the foundation on which family formation is based — people will not cohabit, let alone marry if one of the partners is infertile.

Sex is natural. Sometimes called "nature", it is regarded as an urge that demands satisfaction. That's the way it is. Consequently, sexual activity is considered healthy, repression unhealthy. At the same time, restraint is sometimes necessary, for too much sex can produce ill health. For men, sexual intercourse is a good thing, while for women, it is sexual intercourse linked to childbearing that is good.

Thus, asceticism is of no value in this culture, unlike others such as India and Europe. Except for the brief and symbolic excursion into celibacy by some Rastafari, which in any case is not a ritual prerequisite for membership, Jamaicans do not believe it either possible, desirable or of any special worth to maintain a permanent condition of sexual abstinence. Hence their widespread scepticism towards the celibacy of the Roman Catholic priestly and convent life. On the contrary, they believe such abstinence to be quite harmful, antisocial and anti-nature. Which is to say, anti- the purposes of God.

The naturality of sex erases feelings of guilt and gives rise to early sexual activity. Sex is independent of marriage. Here, it is expected that if two people are in love and are responsible they will have sex, whether or not they are legally married. Among certain cultures, sex before marriage is an unforgivable crime. In the Caribbean, marriage before sex is unjustified folly. This culture places greater value on sexual experience than on virginity.

In this respect, mutual sexual knowledge is seen as a prerequisite to a deeper, more stable and lasting union. It is after and through intimate, including sexual relations that most people take the decision to "make life together", that is to approach the world, its opportunities and risks as a single unit. The visiting and even the common-law unions function as trial forms of marriage. This explains their short life.

Sex is independent of love. One does not have to be in love to have sex. It is an enjoyable activity in its own right, provided propriety is observed. But, on the other hand, love is expressed in sexual intercourse as well as in other ways.

Human reproduction is a function of adult sexuality. Both womanhood and manhood are fully achieved not by the act of intercourse but by reproduction. For the woman, pregnancy and childbirth are the fulfilment of womanhood; for the man impregnation is the proof of manhood. Impregnation and childbirth take on compelling force. Women who postpone this in order to pursue their careers generally consider it a sacrifice.

Turning to the family forms, one of the most important values is placed on children as children. Children are loved for being children uniformly throughout the Caribbean. A child seldom carries the debility of its parents, least of all its illegitimate birth.

Every child has a father. The biological bond is permanent and unbreakable. Even though a father does not take care of his child, he remains its father.

In child care, the role of the father is to "mind" his children, the role of the mother is to "care". This gender role segregation leads to the mother-centredness of the family and the apparent marginality of the father. However, to view the situation thus is to be misled. The father is like God the Father: unseen, definitely present, responsible for sustaining life, the ultimate source of domestic authority and appeal.

The concept of family extends not only to parents and children, but to other relatives. The operative principle here is blood. Blood makes it possible for cousins and sisters, nieces and nephews, grandparents and grandchildren to be appropriated in the family. The nuclear family is not an ideal. The family plot is open not only to one's parents and siblings, but also to one's cousins, aunts and other blood relations, by admission though not by right.

CONCLUSION

These are only some of the values I have time to talk about. My main point is that indeed this is the Caribbean reality we have to live with. This is the reality that has made our people survive, and not only survive, but create; and not only create, but impose their creativity on the rest of the world by its freshness and wholesomeness. Europeans, who have dominated the Caribbean from the first time they peopled these islands with the Africans have generally failed to come to grips with this. They find it very difficult to accept that the Africans in the Caribbean have a culture, let alone one that can provide the foundation for a civilization. Hence their missionary thinking. Whether French, English, Spanish or Dutch, the same thinking is pervasive: these are people to civilize. Hence the imposition of the Columbus centenary.

The church, itself so very much an outgrowth of European civilization, has up to now followed suit, retaining an essentially missionary outlook towards the Caribbean. But it urgently needs to learn to respect the culture of the region and to lose itself in it. Otherwise it cannot find itself.

10

DAILY BIBLE STUDY (3):
THE LIBERATED COMMUNITY AS
SUBVERSIVE AND LIBERATING IN
AN UNJUST SOCIETY

Burchell Taylor

It is clear that Paul's letter, which superficially appears to deal with matters essentially of a personal nature between Paul and Philemon, with Onesimus being the common object of their concern, contains a deeper meaning. Onesimus, we have seen, was more than simply a voiceless and powerless runaway slave who became a Christian convert and whom the law required to be returned to his master by anyone who became associated with him. He, by running away, became initiator of a process that took on significance beyond his own personal welfare and fate.

Too often the church community that existed in Philemon's household is overlooked in reflection on the letter and its meaning. When this is done, it limits the horizon of meaning and the potential significance of the whole experience. The believing community that provided the immediate framework for the practice of the faith cannot be ignored. It is hard to separate Philemon, Onesimus and Paul from that community. Once the liberating impulses that emerged in the whole process are at work and are followed up, the community becomes an essential factor. True liberating insights and possibilities from the Christian perspective, and in response to the challenge and claims of the faith, cannot remain wholly on the level of individual responsibility and experience in isolation from the framework of significant corporate or collective relationships. In this situation, the church in the house of Philemon must be seen as a significant community for the critical experiences of the lives of individual members.

We are, therefore, taking the line that though the letter is personal, it is by no means private. Though the focus is on an individual, it has profound implications for the immediate believing community and, ultimately, for the wider community. One even thinks that the tone of the letter itself gives the impression that it was an official piece of correspondence with the community in mind. It is also true that the greetings extended included others and the descriptive terms used in relation to the others who are greeted bear community reference: "Sister Apphia", "fellow-soldier Archippus" and of course, "fellow-worker Philemon". The senders of the greetings are all termed "fellow-workers". The recurrence of the family term "brother" and Paul's reference to himself as spiritual father reinforce the communitarian significance. There is a love ethic for which Philemon is commended. Then there is the very powerful statement in Paul's reference to the fellowship shared with them as believers which he hoped would become the basis of increasing understanding of the blessings they shared in Christ. This is indeed based upon the reading offered by the *Good News Bible* of a notoriously difficult verse six.

The point, then, is that the return of Onesimus is not a matter of challenge to Philemon alone but also to the Church community. The liberating demand, opportunities, possibilities and risks are shared and will be tested immediately within the community. The barriers that are to come down must come down not only in the personal relationship between Philemon and Onesimus but within the whole community. They must do so right across the board. The benefits of the slave culture shared by members, even as they related specifically to Onesimus, himself a slave, became a factor in the new situation. Would the church be prepared to follow through on this challenge that was emerging for its collective conscience? Would it be prepared to put into effect the logic of the requests accompanying the return of Onesimus?

Paul, under the constraint of all that happened, precipitated by this slave's quest for freedom and deepened by his conversion, asks, among other things, that Onesimus be received not as a slave but as a brother. The revolutionary implications, not only for Philemon but for the church and for the wider community, must not be underrated. Such a request is breaching the socially conditioned understanding they had of the slave and their own self-understanding in relation to him. It was asking for a radical shift in perception of the other that would move him from the periphery to the centre in terms of his designated status and all its social and psychological implications.

Transformation of the community itself is involved in receiving Onesimus. This will mean its own liberation taking place. It will have to accept Onesimus as a brother which brings about a sense of basic and fundamental equality, an equality that is inevitably violated and denied by any practice of slavery or form of oppression. A human being has to be redefined and given a reduced status, to be treated as a slave, or to be oppressed and exploited. This redefinition inevitably gives the edge to the oppressor, as superior in one form or another. This is a superiority which gives rights sanctioned by the status quo. It supplies the basis that legitimizes the situation that puts other human beings at a disadvantage. This is often taken for granted by those who benefit from it.

It is an awakened consciousness by a liberating challenge that often creates the new possibilities. The acceptance of a returning Onesimus on the basis implored by Paul would constitute a direct challenge and offer such possibilities. It would mean the reordering of their own self-perception stimulated by a new perception of Onesimus and, no doubt, others like him. The reordering of this perception would inevitably mean a reordering of attitudes, actions and practices — all in line with the new perceptions. A similar effect would have been created by the request to see Onesimus as representing Paul and that he be received as Paul (17).

What is being put forward is a reinforcement of the liberating equality. When it is hinted that he is no longer useless but useful, it is a challenge to redefine human worth in terms of human dignity as a human being (11). This further lays the basis for new liberating action and practice. Giving a human being full human worth contradicts seeing and using that same human being as a slave, however benevolent the institution may be made.

Barriers within the believing community must come down as perceptions and practices are radically reordered. The liberation of the community itself is essential. It was not Philemon alone who was involved. Indeed, he needed the community's support as much as the community needed his courageous lead for the whole project to be implemented. It is the community that would provide the framework for full effect of the new situation and condition. It would provide the necessary and immediate framework for the outworking of the liberating practice prompted by the acceptance of Onesimus on the basis suggested. It would provide the framework for the practice of love for which Philemon had been well known and for even more meaningful expression of the fellowship that is shared between Paul and themselves.

There would be a new and deepened sense of solidarity in an important and far-reaching liberating project and to which reference has already been made. Surely the liberating challenge becomes a test to real love and fellowship. The Christian community becomes the framework for effective realization and fulfilment of the challenge. This is the nature of a liberated community. There is seen here the radical reformation, better yet, transformation of a community whose religious practices, in terms of devotional and liturgical practices are being confronted by and linked to practice of ethical and social significance. It means accepting and relating to another human being on the basis of the integrity and dignity of the person's whole humanity. A believing community that truly does this will find it impossible to sanction its contradiction in the wider community or benefit from such a contradiction.

No doubt, despite the difference of opinion that surrounds the meaning and implications of Paul's declaration of there being no difference between Jew and Gentile, male and female, bond and free (Gal. 3:28), what is expected of the church community in Philemon's house is nothing but its logical, practical demonstration and outworking. It is a radical reordering of relationships and status that defies and contradicts current social arrangements which were taken for granted as conforming to how things were always meant to be. It is a new liberative factor introduced into the process in light of the impact of the Onesimus-Paul encounter and experience.

If the community of believers in Philemon's household accepted the liberating challenge that receiving Onesimus as a brother and not as a slave would constitute, then, as a liberated community itself, it would, in a real way, also become a subversive community and at the same time a liberating community, in its given context. Its own practice and presence now in a wider community that existed on the basis of structures of domination and dependence, and oppression and exploitation would offer an alternative perspective and practice which would be contrary to the accepted pattern for which it was generally agreed there was no alternative. The church would be the same small community meeting in the household of Philemon. Its social significance otherwise would have been minimal, bearing in mind the fledgling nature of the whole Christian movement then. It would not then launch out on any great vociferous crusading campaign to challenge existing structures in society.

Nevertheless, with the barriers within its own fellowship collapsing and with a radically new relationship implemented that affirmed basic human equality and dignity, there was in the making a direct challenge to the way things were in the wider community.

All too often this is where and how the liberating process begins. It begins where the community of believers puts itself at risk by being true to itself in light of an awakened consciousness. It is a consciousness awakened not by self-generated charitable feelings but by being confronted by the oppressed becoming challenging resource persons in light of their own experience and understanding of God's gracious liberating power. This is what a returning Onesimus would become. He himself would be returning at great risk to himself, but motivated by the experience and understanding referred to, makes such a return necessary, whatever else might have been at stake.

A liberated community that becomes subversive by its presence and practice in an unjust social order will, therefore, have to come to terms within its own ordered life, with such essential liberating realities as equality (4,8,16,17); solidarity (12, 13, 17); human dignity and worth (11); love as an essential outworking of faith (5, 7, 9); fellowship in Christ (6); going the extra mile as a freely accepted imperative of faith-commitment rather than as a concession of pity (8-9, 21). These things do not exhaust essential liberating realities but they go a far way in pointing to important factors that are necessary. Where they are present as part of the current practice, the liberating experience is at hand.

- The church becomes and remains conformist when it reproduces in its own life and pattern of relationships similar structures and orders of relationships that prevail in an unjust society. It does not give itself any real opportunity to become an effective, liberating presence in relation to the whole socio-political reality in this regard.

- The church underrates its liberating effectiveness when it links such effectiveness to size, power and material resources. It overlooks the power of the subversive nature of its own life which, if properly ordered, could be in the midst of a community of injustice and oppression.

- The church puts itself at necessary risk when it becomes liberated itself, and this becomes a challenge to the existing unjust order by offering an alternative perspective and practice to what is often presented and justified on the basis of there being no alternative. It is often this practice that brings about persecution and martyrdom.

Probably a more sensitive reading of this letter of Philemon, instead of irritating us by what appears to be concessions it made to an unacceptable state of affairs in the ordering of human life, would challenge us concerning the liberating potential of the life of the believing community itself. If this challenge could lead us to the self-criticism that is an essential part of repentance, this would be very good for us in our context. Repentance emerging from such challenge opens the door to the renewal that is important for liberating promise and purpose.

11

MINISTRY FORMATION FOR THE CARIBBEAN

Howard Gregory

INTRODUCTION

In attempting to explore the topic which has been assigned to me I would like to lay my cards before you in order to clarify some of the parameters which have defined my approach to the task.

I understand ministerial formation as a specific area of focus or concentration within the broader enterprise called theological education. To that extent I would regard ministerial formation almost as a sub-discipline within theological education.

For the purpose of this paper I shall use the term "Caribbean" to refer to the English-speaking Caribbean, which I know best, while at the same time acknowledging my indebtedness to insights which have been gleaned from the Spanish-speaking areas. I do so also cognizant of the Latin attempt to redefine the region from within in terms of Abya Yala.

I shall approach the subject with a bias as a pastoral counsellor. As such, I have a special concern about the personhood of those who participate in the formation process and a vision of wholeness for those who share and those who receive ministry in a way that takes seriously the psychological perspective.

As the Chief Executive Officer of the United Theological College of the West Indies (UTCWI), charged with the double responsibility of ministerial formation and theological education, I have an interest which goes beyond the denominational concerns of participating communions. It is my task to ensure that the institution operates in a way consistent with notions of education in the worlds of academic and professional development. It is also my responsibility to ensure that the institution operates in a way consistent with sound administrative principles, so being stewards of the resources which have been entrusted to us. To that extent, I may be accused of being a bureaucrat in my approach to some issues.

In approaching my task I am conscious that there is a history of ministerial formation in the Caribbean which has seen the formation of ministers who have served with distinction in the Caribbean and various parts of the world. While some may be tempted to see in this a failure to train persons for cultural particularity, I am inclined to see in some of these comments an ongoing denigration of ourselves in which we fail to acknowledge our people and our achievements.

MINISTERIAL FORMATION AND THEOLOGICAL EDUCATION

As a starting point for our reflection, I would like to draw some distinctions between theology, theological education and ministerial formation. Here I am indebted to the works of Julio de Santa Ana and John Pobee.

Julio de Santa Ana dispels the notion of theology and theological education as synonyms. Theology is perceived as the articulation by human beings of their experience of the revelation of God. In this way, theology begins not with discourse but with human experience. It is this notion that theology is the product of this two-step process that has led Gustavo Guitierrez to speak of theology as a "second act".

It would, however, be misleading to think that theology consists of a body of answers. Theology may equally be conceived of as questions. Put another way, that one may even dare to ask questions about God and about human experience, may indeed be an expression of theology. All these elements, the experience, the questions, the answers, go on at an individual and corporate level. In the history of the community of faith, theology has come to represent not just the articulation of the new or idiosyncratic, but has served to "confirm", "correct", "deepen" or "restate" earlier articulations of experience.

Theological education is inextricably bound up with the life of the community of faith. It involves training in the answers and methodology of the "second act". In that way theological education involves rehearsing the answers of the past, while at the same time asking and engaging the questions of the present within a context of historical and cultural particularity. According to Julio de Santa Ana, "theological education is offered to facilitate theological production and make the latter as relevant as possible: relevant to the community's faith; relevant to the community's traditions and also relevant to the situation in which the community is living".[1]

In our contemporary context, theological education has been used not only as a synonym for ministerial formation, but to speak of the offerings of the seminary or theological college. This represents a distortion of the reality. Theological education is something which can and ought to go on at the most basic level of Christian community and in the formal academic setting of the seminary. In this regard the Latin American experience of Theological Education by Extension (TEE), especially that at Seminario Biblico, is clearly demonstrating that participation in theological education requires no formal matriculation in terms of formal academic credits. It not only trains grassroots leadership for the church but it keeps them grounded in the life of the local community.

At another level there is need for more vigorous and formal education of those who direct the educational process at various levels within the life of the church and those who are charged with the task of giving articulate, coherent, and systematic expression to the faith within the context of the church. The history of the church is replete with such persons and institutions who have served to articulate the faith and to protect it from distortions. It is in this arena that the formal education of the seminary or theological faculty, with its matriculation requirements, normally fits.

For some persons it is precisely here that they would fault today's seminary or theological college, in that its perceived focus on vigorous academic preparation is not serving the needs of the church. It is here that we come to appreciate the distinction between ministerial formation and theological education.

Not all theological education is ministerial formation. Ministerial formation includes theological education and has a focus which takes it beyond these borders. As John Pobee writes of ministry formation: "it is about . . . the moulding by discipline, instruction and organization of persons and institutions so that they may be true embodiments and servants of what ministry is all about".[2]

Captured in this definition are two paradoxical relationships identified by Pobee. First, that between sound theology and lifestyle, and secondly, that between knowledge and praxis. To these I would add a third, vocational and personal identity. Seen in this light, then, theological education precedes and provides the context within which ministerial formation takes place, while ministerial formation represents an intentional structured focus beyond theological education.

MINISTERIAL FORMATION IN THE CARIBBEAN: A HISTORICAL REVIEW

In looking at the history of ministerial formation in the Caribbean, it must be acknowledged that there is little evidence of a distinction between theological education and ministerial formation. A rather long history of theological education traces its roots to the post-Emancipation period. These early attempts were intended to prepare "native" leaders to work among the ex-slaves population in rural communities. The congregations in the towns would continue to be served by missionaries from overseas and these would also provide leadership for the national churches.

Horace Russell identifies the motivating force behind the birth of theological education in the Caribbean as the provision of:

- the Christian leadership needed for the creation of a new society;
- the leadership needed in response to an awakening to things African in the peasant population and to develop a missionary thrust to the African continent;
- native replacement personnel for the European leadership which was now migrating elsewhere.[3]

The earliest attempts at theological education represented denominational initiative. Clement Gayle identifies several of these in the following terms: "Such attempts were made by the Anglicans in Barbados and Guyana, the Baptists and Presbyterians in Jamaica, the Lutherans in Guyana, the Moravians and later the Anglicans in Jamaica."[4]

Earlier reference to the development of theological education in the Caribbean indicated that it involved a process of ministerial formation intended to provide pastors for the "native" population of ex-slaves in rural communitites. No doubt this early training had a very functional orientation.

Over time, the denominational college provided a course of study which was very much along the line of the classical model. Prior to the formation of the UTCWI, most students pursued a diploma offered by the different colleges. In the case of the Anglican students at St Peter's College, they pursued the General Certificate of Ordinations Examinations which were set in England and set for candidates for ordination through the Church of England. A select few from the antecedent college were allowed to pursue the London Bachelor of Divinity (external) degree. In time it was to this small group that the UTCWI looked for its Caribbean faculty members.

The birth of UTCWI marked a new phase in theological education in the Caribbean as the academic standards and opportunities were widened to embrace a greater number of students. The Licentiate and the Bachelor of Arts (Theology) soon became normative for students. In addition, students were no longer studying only classical theological subjects but were also studying social sciences and other courses offered at the university.

A serious shift took place in ministerial formation at the same time. In the widened ecumenical context, the central core of the College's life was the academic programme along with corporate worship. Much that was considered ministerial preparation was retained under the control of the denominational slots, yet there was no uniformity as to what was considered adequate ministerial formation.

In this context there has arisen constant discussions about spirituality. The College has constantly been criticized for its shortcoming in turning out graduates without an appropriate spirituality, though there is no agreement as to what this should be. It appears that spirituality is at times really a cliché for a desired balance in the formation of ministerial candidates. Has there really been too much emphasis on academics? Have the students lost something vital in the relationship with university? Have they fitted the expectations of the churches?

One problem of course, was that the Caribbean context in which ministerial formation was taking place was also changing. The Black Power movement led to a questioning of some of the structure of authority, including the role of the minister. At the same time, ministerial candidates were also asking about the relevance of church music to Caribbean peoples. In the early 1970s there was a birth of indigenous church music in which the UTCWI students became involved. In addition, there was an interest in the relationship between the arts and religion in the Caribbean context.

Back in 1967 there was a growing awareness of the need for new expressions of ministry in the Caribbean. Accordingly, a project called the "Mandeville Project" was planned to involve faculty, students, a sociologist, and community representatives in an examination of the impact of bauxite mining on the community of Nain. This early attempt ended with the College retreating, as the multinational bauxite company had objections to the project. It represented, however, an attempt to take theological education beyond the walls of the College and to relate it to the life of the community.

As a follow-up to this project, a conference sponsored by the College and the Theological Education Fund entitled "New Forms of Ministry" was held. The main task of the conference centred around the following issues:

- traditional patterns of ministry in the West Indies
- ordained ministers and other ministries
- the UTCWI programme as it related to the first two concerns
- the use of centres throughout the West Indies for an ongoing programme of theological education for the churches[5]

Among the recommendations of the conference was the need for improved facilities for training in practical fields such as:

- administration of local churches
- training of lay people at the congregational levels
- various chaplaincies (schools, hospitals, prisons)
- counselling, including marriage counselling
- stewardship, including the proper management of personal and church finances; perhaps a course in bookkeeping
- home nursing and first aid
- communication, not only through speech training, also through the use of mass media: the press, radio and television
- the use of group dynamics
- musical training, and perhaps some training in the fine arts, and
- school management for those who may be called to render this service[6]

Influenced by developments in clinical pastoral education in the United States and the entry to the faculty of persons trained in these disciplines, the College began to widen its offerings to include courses in pastoral care and field education. The field education programme was intended to provide supervised learning experiences for those in the process of ministerial formation while working in a context of human suffering and crises. This represented an attempt by the College to take more direct control of the formation process which, up to this point, was controlled totally by the churches. In recent years, this has been widened to include children's homes, homes for the aged, the prisons and other community-based projects.

While these various emphases have arisen at various stages in the development of theological education in the Caribbean and in the College in particular, it does appear that the task of integration is still to be achieved.

The Call

Ministerial formation begins with a personal experience of a sense of one being "called" to fulfil a particular vocation. This sense of call may be dramatic and sudden and be accompanied by a sense of urgency. It may, on the other hand, be a long, slow, and undiscerned process for a while. It may even be accompanied by moments of struggle and turmoil in the life of the individual as he or she attempts to resist this sense of calling. In either case, the moment of revelation may consist of a conviction of God laying hands, as it were, upon the person, or a deep awareness of human need and suffering which demands a commitment to caring.

In a Caribbean church facing a shortage of candidates for the ordained ministry there needs to be a clear understanding of the nature of the call to ministry in order to determine whether we wait for vocations or we take initiatives which bring persons face to face with the possibility of having a vocation to ministry.

Ministry Formation is Grounded in the Church

It is not enough to simply have an experience of a call to ministry, what has been termed the "immediate call". Ministry belongs to the church, not as institutions, but as the body of Christ. It is therefore necessary for those who claim to have the experience of a "call" to present themselves to the church to be tested as to their calling and whether they possess the qualities requisite for the exercise of ministry. Both elements are essential in thinking about ministry. As Thomas Oden expresses it: "It is incorrect to say either that one is properly called by God alone without the church, or properly called by the church alone without God."[7]

Ecclesiology and Missiology

Ministry formation is not just a matter which concerns the individuals who claim to have experienced a call from God, the judicatory authorities of denominations or the training institutions themselves. Ministry formation is in the service of the whole church. Ministry is part of God's will for his church and a demonstration of the community's exercise of true discipleship to Jesus Christ. The ministry of the church is, in fact, an extension of the ministry of Jesus Christ into which we as Christians are called to share. Ministry in its diversity is only a gift of Christ that serves the church, the body of Christ, and people of God.[8]

The Ministry of the Whole Church and the Ministry of the Ordained

In attempting to get feedback about the programme and levels of participation in the Lay Institute of the College, we have received reports that some lay persons are discouraged from doing the courses we offer by their parish ministers, on the premise that they do not need to do these courses. Further exploration indicates a sense of threat that lay persons should have the veil of mystery removed from around some things related to the faith. Needless to say, this is the beginning of an experience of frustration and conflict for lay persons who proceed to undertake the courses.

There exists a serious need for clarification of the relationship of the ministry of the whole church to that of the ministry of the ordained. Ministry is entrusted to the whole church by virture of their baptismal incorporation into Christ. This has been expressed in terms of the New Testament idea of the priesthood of all believers. The challenge for the Caribbean church is to discover whether there is an appropriate model or approach to ministry formation which is geared towards the facilitation of the priesthood of all believers.

We need to note, however, that the notion of the priesthood of all believers is not the end of the story. Specific spheres of ministry (or ministry functions) have always been entrusted to representative persons who act on behalf of the whole faith community. A creative tension needs to be maintained between the two. As Oden writes:

The office of care of souls is given to some few who are called to represent the whole Church in the crucial situations of preaching, sacramental administration, church order and pastoral care. The sacred ministry is not to be affirmed in such a way that the ministry of the laity is neglected or denied, but affirmed and enhanced. On the other hand, the general ministy of the whole Church is not to be asserted in such a way that the ordained ministry is disenfranchised or diminished.[9]

Challenges for Ministry Formation in the Caribbean

One of the first challenges facing Caribbean churches involved in ministerial formation is that of determining the nature of the ministry for which one is being prepared. This is important if the experience is to prove one of enhancement for the ministry of the individual and that of the whole church and not be a venture in mutual frustration.

This calls for determination of whether the term ministerial formation has anything to do with the general ministry of the whole church or is confined to that of ordained persons. This introduces the element of diversity in the ministry to which persons are called. This stands in contrast to much of what has transpired up to this point. For a long time, ministerial formation has been guided by a model designed to serve single males living in a residential institution. This is at times condescending to female students and, in some quarters, when ministerial students are being counted, the female students are treated as an addendum. Also married men and their families have, at times, been sidelined in the formation process, and female students who later marry and have families are an irritant to denominations who see the resulting complications as their problem which they need to resolve and then report to the denomination.

Whether male or female, however, there is a high level of frustration among many students who, from the outset or during their pilgrimage, have doubts as to whether or not parish ministry is that to which they are called. Sooner or later these persons have to fall in line and fit the mould as denominational funding of lay persons for lay ministries is rare, and preparation for specialized ministries or pioneering ministries is even more of a rarity.

Not only do the churches need to give serious thought to the handling of diversity within the ministries of ordained and lay persons, but theological colleges and seminaries must also look at what preparation they offer their students for understanding the relationship between these two segments of the church. Models of preparation have generally encouraged the notion of the ordained as the one who single-handedly manages the ministry of the church without imparting ways in which to empower, enable, motivate and cooperate with lay persons in the exercise of the ministry of the church. There is, therefore, too much evidence of a sense of competition, diminishment and threat in the perspective of students and graduates. Course offerings in ministry need to be reviewed and upgraded in this regard.

Individuality

It has already been observed that ministry formation begins with the experience of a sense of "call" on the part of a prospective candidate. It has also been noted that diversity in ministry needs to be acknowledged if the church and training institutions are to give authentic expression to the call of each individual. It seems to follow that there is need for an approach to ministry formation which facilitates the exploration of selfhood in order to help the candidate to discover himself or herself and what he or she brings to ministry by way of a personal story.

Anyone involved in ministry formation in the seminary or theological college and who has attended homiletics classes or listened to student sermons would witness repeated attempts by students to ape outstanding preachers within their denomination. While there is nothing wrong with having "idols", these should inform the height to which one strives and not set one on a path to become a carbon copy of the other. Carbon copies are lacking where authenticity is concerned.

Daniel Day Williams places the whole matter of self-knowledge within the context of ministerial formation thus: "In a Christian view, self-knowledge has three dimensions: it is theological, for the self is God's creature; it is personal, for each self is a unique centre of experience, and it is vocational, because every self has something to do in the world which requires his special contribution and personal decision."[10]

Many models of ministry formation place the ordained minister over against the lay persons he or she seeks to serve. Ministry is, however, an interpersonal vocation which calls for self-knowledge and the celebration of one's personhood with all its aches and pains. The dangers inherent in approaches which seek to ignore this dynamic are highlighted by Williams, "Physician, heal thyself! Those who undertake the care of souls must attain self-understanding. We have seen how the counsellor's inner life is involved in his healing ministry. The pastor can obstruct the work of grace if he does not understand himself or his people."[11]

One threat to this kind of self-knowledge is inherent in much of the process of preparation of Caribbean pastors. In view of the shortage of pastors, there is a tendency for ministers in training to be given positions of responsibility in which they perform ministerial functions with little supervision, and especially the kind of supervision which involves introspective work. The minister in training may have no difficulty with this model as it gives a sense of importance to one who is able to exercise such a ministry. Yet this strategy prematurely creates vocational identity without exploring the underlay of personal identity. This, therefore, produces a functionary who has been denied some of the possibilities for personal growth of which Rosario Battle speaks when she says: "Learning should be a transforming activity whereby the learner's values are changed, self-knowledge is increased, and becoming is activated."[12]

Churches and seminaries will need to enter into a new kind of relationship in which more opportunities for supervised clinical experiences are offered while intramural training lasts. These experiences can be organized either by the seminary or in cooperation with the churches. At the same time the seminary needs to assist the churches in developing a cadre of persons involved in ministry supervision. It is a fallacy to assume that every pastor who has experience in ministry is suitable as a supervisor of ministers in training. I have worked with ministers of almost every denomination represented in the College while they supervise students and found that some of these experiences border on what may be called a disaster. The minister in training and the churches have everything to gain from a well organized supervisory training experience which facilitates personal and vocational identity and growth on the part of those in training.

INTRAMURAL, EXTRAMURAL OR THEOLOGICAL EDUCATION BY EXTENSION

A document prepared by the Evangelical Faculty of Theology (FEET) in Managua, Nicaragua offers a twofold classification for designing theological education. It says "Theological education can be designed according to two extreme models:

a) the elitist model, a group of students enjoying full scholarships, and
b) the popular model, an extension programme which offers courses

for individual study based on workbooks financed by the institutions."[13]

Against this background there is hardly a need to debate the nature of the prevailing model for ministerial formation in the Caribbean. Indeed, it is obvious to me each year when the budget has to go to the Board.

Not only is the model a very costly one but it effectively removes the candidates for ministry from the local community and leads to a process of deculturation by which students may become alienated from the local communities and families within which they have been nurtured. This is particularly true of those who come from rural communitites.

In addition, the residential model is one which tends to be strongly focused around the degree-granting programmes. While I am fully committed to the value of such a programme, I have doubts about its usefulness for everyone. Many persons who eventually enter such a programme come to college without matriculation requirements for entry initially. I am of the opinion that some of the criticisms of stress, inability to integrate and lack of attention to pastoral courses result from the struggle of some to simply keep up with the rigours of the academic content and structure of the programme. In this context the academic nature of the ministerial formation process is likely to be blown out of proportion.

One participating communion, the United Church in Jamaica and the Cayman Islands has sought to develop an alternative model for ministerial formation, the Institute for Theological Leadership Development (ITLD) programme. The experience of this programme has not yet been made public.

Reports coming out of Costa Rica, where the TEE model has been used with great effectiveness, indicate that there is a serious cost factor involved in taking faculty to students in their various settings. Indeed, my own professional training as a teacher was undertaken using a similar model for aspects of the programme and it was discontinued by the UWI because of the cost involved. The development of communications technology is making this approach possible in some areas of the teaching life of the UWI, yet even here the cost for use of the distance teaching facility (UWIDITE) is very high.

It seems to me that what is needed is an approach to ministerial formation which seeks to bridge the gap between formal intramural programmes and the communities in which people live their daily lives. This would involve placement in outreach projects of churches and in rural communities for specific periods for purposes of ministry and reflection. This would be similar to models used in the training of various professionals including doctors, nurses, teachers, and so on.

One concern about TEE models for ministerial formation of full-time ordained ministers is that, while they facilitate community-related training, they run the risk of creating a second-class group of ministers. Perhaps it is too early in the day to tell, but it will be interesting to see what pastoral charges are entrusted to those trained in such programmes.

These models do not in themselves resolve the issue of whether ministerial formation concerns the ministry of the whole church or whether it involves the ministry of the few to be ordained. This College has recently embarked on a programme which is intended to facilitate lay persons who seek to acquire formal theological education in order to enhance the various lay ministries in which they are currently involved. The College's timetable is structured to allow them to attend early morning and late evening classes on a part-time basis. This is an area that has great potential for growth. The challenge will be to take this to a wider constituency in more remote places.

THEOLOGICAL EDUCATION AND MINISTERIAL FORMATION

It has already been observed that theology represents the articulation of the answers and even the questions of persons based on their experience of God in the particularity of their context. Theological education has been defined as the training of persons to help people to search for and to find answers to the theological questions which they raise.

The question which we need to ask is this: With that theological starting point do we begin the process of theological education leading to ministerial formation in the Caribbean? It appears that too much of the basic theological elements begin with theological answers which are not related to the questions our people are asking in their context. Accordingly, they may represent sound articulations of systematic theologies from other contexts but not from that of our Caribbean peoples. This is indeed the challenge, not just for the professors of systematic theology but the entire range of offerings in theological education.

Ministerial formation which is relevant to the Caribbean and faithful to the mission of Christ must begin with theological reflection which is grounded in our understanding of the realities of the context in which we live. Only then will the ministers be able to be the prophets, mouthpieces and agents of God, who challenge people with the demands of God and seek to make God a reality in the life and experience of the peoples of the region.

The realities of the Caribbean will be covered elsewhere, but I need only point out that this region is one characterized by the following:

- A stranglehold of dependency — energy dependence, dependence on Western interests for the export of primary commodities — and financial dependence on Western loan flows.

- The problem of debt and the associated problem of the cross-conditionality of lending agencies. The demand of lending agencies for the three Ds — devaluation, deregulation and divestment with serious negative social and economic consequences for the people of these lands.

- A general mood of impotence and hopelessness in face of the serious reductions in social services, education and health care.

Professor Franz Hinkelammert has characterized the developing countries on this region as suffering from a "culture of despair".[14] Driven by a neo-liberal philosophy of the marketplace, there develops a growing sense of anonymity among our peoples, the breakdown of social relations, the emergence of the drug culture — an economic strategy guided by a thesis that there is no alternative to the path being pushed at us — and a gnawing sense of despair with less opposition because of the difficulty justifying one's opposition.

If theology and the theological education offered to those in formation for ministry are to be adequate, then they must be grounded in these realities in which the questions and answers of people must arise. It is here that the TEE model of Costa Rica has a lot to offer us. It seeks to create an understanding of ministry formation as involving the facilitation of the search for meaning by lay persons as they ask their questions about God's involvement in the midst of this reality of life and their world. Our ministry formation seeks to provide people with answers, even to questions they are not asking, rather than facilitate their questioning.

Maintenance versus Transformation

This College and its participating communions often dialogue about the ministers whom we send to the churches. There is regular criticism that we are not providing adequately trained ministers for the local churches. I am quite baffled at times as to what this means, particularly so when it comes with the suggestion that the antecedent colleges did a better job. It seems to me that there is a feeling somewhere that if we only do now what was done then, then all would be well. I submit that that is a fallacy.

I believe, however, that what is being stated in these criticisms is that the graduates do not perform satisfactorily in the maintenance functions of the local congregation. While appreciating the fact that churches need maintenance-oriented personnel, the question arises, to what extent is the educational process intended to serve a maintenance function versus a transformative one? Are the ministers to be agents of transformation of the life of the peoples, the communities, the congregations and the churches? If theological education is truly educational, then it will, of necessity, produce some round pegs which do not fit square holes. This, of course, is a matter for dialogue.

Rethinking Ministry

It has already been observed that the predominant notion of ministry formation as concerned with the training of single males for parish ministry is losing its usefulness and validity. In this regard, the church needs to come to terms with the ministry of women. It is folly to assume that a synod vote to determine the ordination of women settles the issue. The increasing number of females in seminary, the tremendous leadership role being played by women in local churches, and the increasing marginalization of males in a context like Jamaica, mean that the centres of ministerial formation must encourage dialogue, positive attitudes, and the creation of models for women in the ordained and lay ministries.

In addition, the church and seminary need to see ministerial formation as something concerned not just with a single-mould ministry, namely parish ministry. John Pobee identifies four ministries for which those involved in ministry formation must prepare persons:

- those who will be parish priests or circuit ministers who represent a kind of general practitioner
- those who are going to be academics who, nevertheless, cannot be freelance theologians but articulate theologians in the womb of the church

- those who are going to be in the specialized ministry of communication
- counselling ministry[15]

To this list I would add and highlight certain categories:

- Ministers of youth
- Ministers to the aged
- Ministers for rural settings
- Ministers for inner city settings
- Ministers for community development and mobilization

ECUMENICAL VERSUS DENOMINATIONAL APPROACHES TO MINISTERIAL FORMATION

One question that persons from outside the ecumenical context represented in the College often ask is: How do you all survive together? People seem to have images of major eruptions in the life of the institution centred around denominational differences. Elements closely associated with the institution often propose denominational strategies or retreat to a state of denominational nostalgia when confronted by difficult issues in the life of the institution — spirituality, ministerial formation and, believe it or not, the cost of funding the institution.

Development in ecumenical theological education in the Caribbean has had a long history and did not come easily. From very early there were theological bases for ecumenical cooperation in theological education. It was recognized that the bridging of denominational barriers was expressive of the unity of the Body of Christ for which our Lord prayed. Further, it represented a reasonable approach to the stewardship of the resources of various branches of the church in the Caribbean.

A major milestone in this process was marked about 1961 when the West Indies Federation fell apart. The various churches which came together to form the UTCWI determined that the churches should be a vehicle for the unification and development of Caribbean society and that this task would best be undertaken as an ecumenical effort.

It is clear that there are still denominational colleges for ministerial formation. Yet it is equally clear that the ecumenical commitment which funded this College has changed the face of theological education and ministerial formation in the Caribbean. In real terms, it is true to say that every Caribbean theological institution represented at this

Consultation is ecumenical. Few parts of the world can understand the relationship between St John Vianney, Codrington, UTCWI and St Michael's. St Andrew's was ecumenical before UTCWI came into being, and Templeton Seminary has been ecumenical from its inception.

The retreat to denominational shelter or cloister is merely an attempt to avoid the painful work of searching and struggling with the issues of today. It is an attempt to return to the good and safe old days which, perhaps, never existed. But more importantly, it is an attempt to renege on the mission of the churches to Caribbean peoples and to the cooperative endeavour which is expressive of the Body of Christ.

Spirituality and Ministry Formation

It seems that whenever the issue of ministry formation is raised, someone is bound to start talking about spirituality and the failure of the seminary to train persons in spirituality. This, of course, is an item for controversy in an ecumenical institution as there is no consensus as to what is meant by spirituality and how one imparts it. Any attempt which I shall make to comment on spirituality will, therefore, be subject to the same scrutiny and criticism.

Samuel Amirtham and Robin Pryor speak of spiritual formation in these terms: "Spiritual formation is both the process of all of life, and also the very specific, planned experiences in which we engage the thinking — praying — interrelating of our theological students, faculties, congregations and wider communities."[16]

This statement seeks to express the notion that spirituality is not just a little compartment of life but concerns all of life and at the same time has intentionality and must be expressed in specific activity.

Unfortunately, students come to the College with little understanding of spirituality beyond some of the narrow neo-Pentecostal definitions prevalent in our society, and we do a less than admirable job in helping students to develop a spirituality and to grow. At the same time we need to be aware that some of the denominational responses concerning a desired spirituality are also lacking.

John Pobee helps us to understand some groundwork for addressing the issue of spirituality. He first of all offers a definition of spirituality from M. M. Thomas:

> The structure of ultimate meaning and sacredness within which human beings live and enter into a relationship with nature and fellow human beings in politics, economics, society and culture . . .
> The primary concern of the Christian Mission is also with the salvation of

human spirituality, with humankind's right choices in the realm of self-transcendence, and with structures of ultimate meaning and sacredness — not in any pietistic or individualistic isolation, but related to and expressed with the material, social and cultural revolution of our times. The secular strivings for fellow human life should be placed and interpreted in their real relation to ultimate meaning and fulfilment of human life revealed in the divine humanity of the crucified and risen Christ.[17]

He then goes on to lay two caveats to be observed in approaching spirituality for ministerial formation. The first is: "Each spirituality is contextual. We may not uncritically take over the images and ideas of spirituality from another part of the world . . ." It is interesting to note that the Consultation on Spiritual Formation in Theological Education held in Indonesia, 20-25 June 1989 makes the same point:

> As applied to spiritual formation, there is need to liberate spiritual formation from imposed models, e.g. the assumption that contemplative meditation is the acme of spirituality and better than a spirituality of service. Christian spiritual formation must build on the spirituality of the society.
> For that reason, there is need for explorations into local art, history, images, symbols, folklore, dance and drama to serve as tools for expressing and communicating new forms of Christian spirituality.[18]

The second point is that:

> Spirituality has a progressive and changing nature. Therefore, the spirituality that we seek in ministerial formation cannot be only reliving past models of spirituality but also new spiritualities for new times, albeit nurtured by scripture and the examples of Christ. In light of these observations, it seems to me that the call for spirituality in ministerial formation often ignores the need for the contextual work that has to be done. Failing this, we shall simply have an imposition of more of the old and little that is authentic and liberative in the spirituality we seek to impart. The challenge is for seminary and churches to work together on these issues.

In the meanwhile, there are some things that the colleges can do. One of the elements identified as a contributor to the decline of spirituality, especially by students, is the impact of the critical study of the Bible on their long-held spirituality. We are reminded, however, by J. Severino Croatto that this need not be the case. Thus, he writes:

We have insisted that critical study of the Bible and of theology during
theological training, far from driving away spiritual experiences, deepens
and enriches them. It should not produce spiritual emptiness, later to be
filled from other resources. Critical study is an adventure, both in
intelligence and in faith: to the extent to which one discovers who God is,
how he manifests himself in histroy, how he incarnated his historical project
of liberation, so one's faith grows.[19]

He does, however, insist that those who train must know how to
guide. It is, therefore, the obligation of the teachers to assist the process
of spiritual formation by their approach to their teaching and perhaps
by the appreciation of the interconnectedness of spirituality and critical
thinking.

Further responsibilities may be entrusted to the teachers of
ministers in training in terms of modelling an appropriate spirituality
and in structuring opportunities and the environment for the nurture
and growth of spirituality, especially lifestyles of prayer. Six such
elements are identified by the Spiritual Formation Consultation in
Indonesia as follows:

- a course on theology of prayer
- worship opportunities within the school, planned and
  conducted jointly by educators and students
- retreats under the leadership of acknowledged prayerful
  educators and other praying spiritual leaders
- visits to communities of prayer to share in their lifestyle
- setting aside quiet rooms for private prayers
- encouraging students to form and find their own prayer groups
  as an element in the seminary as a whole growing into a praying
  community[20]

CONCLUSION

It is impossible in a paper of this nature to do justice to all the issues
which are raised. One therefore has to exercise some judgement
concerning the most salient issues. Accordingly, I have sought to define
the subject of ministerial formation as a particular area of focus or sub-
discipline of theological education. I have sought also to indicate that
ministerial formation has a history within the Caribbean and is,
therefore, not a new road to be explored.

Recognizing that ministerial formation cannot be discussed in
isolation from persons, the church, and the mission of the church, I
have sought to very briefly introduce these as a context within which

to explore the peculiar Caribbean issues. I have raised a number of challenges and issues which arise in thinking about ministry formation in the Caribbean. I have not been as definitive as some would like and I have probably raised more questions than provided answers. Perhaps it is my way of saying that there is no single strategy that will encapsulate all that is needed for effective ministry formation in the Caribbean, and to indicate that the process is dynamic and needs ongoing reflection and adjustments.

NOTES

1 Julio de Santa Ana, "Theses on theological education". Paper delivered at the Conference on Theological Education in Situations of Bare Survival, 14-18 July 1991, Managua, Nicaragua. Translated by Kield Renato Ling, 2.

2 John S. Pobee, ed., *Ministerial Formation, Mission, Today's World.* A collection of papers from a Consultation on Ministerial Formation for Mission Today, held in Limuru, Kenya, 1989, 4.

3 Horace Russell, "A brief account of the formation of the United Theological College of the West Indies and its development". Mimeo, 1991, 11.

4 Clement Gayle, "History of the United Theological College of the West Indies". Later published in the Twenty-fifth Anniversary Souvenir Magazine of the United Theological College of the West Indies, 1991, 1.

5 Horace Russell, op. cit., 10.

6 Conference Report, 8.

7 Thomas Oden, *Becoming A Minister* (New York: Crossroad, 1987), 32.

8 John Pobee, op. cit., 5.

9 Thomas Oden, op. cit., 81.

10 Daniel Day Williams, *The Minister and the Care of Souls* (New York: Harper & Row, 1961), 96.

11 Ibid., 95.

12 Rosario Battle, "A structure that encloses formation for mission", in *Ministerial Formation, Mission, Today's World*, edited by John S. Pobee, op. cit., 35.

13 A document prepared by the Evangelical Faculty of Theology (FEET), Managua, Nicaragua, and presented at the Conference on Theological Education in Situations of Bare Survival, 14-18 July 1991, Managua. Translated by Kield Renato Ling, 1.

14 Franz Hinkelammert, "Abya Yala and the current world situation". Paper delivered at the Consultation on Theological Education in Abya Yala, 20-24 July 1992, San José, Costa Rica.

15 John Pobee, op. cit., 5.

16 Samuel Amirtham, and Robin Pryor, eds., *Resources for Spiritual Formation in Theological Education* (Geneva: World Council of Churches Programme on Theological Education, 1990), 4.

17 John Pobee, op. cit., 8.

18 Samuel Amirtham, and Robin Pryor, eds., op. cit., 82.

19 Severino Croatto, J. "Spiritual formation and critical study" in *Resources for Spiritual Formation in Theological Education, op. cit., 16.*

20 Samuel Amirtham, and Robin Pryor, eds., op. cit., 85.

12

CONCLUSIONS:
THE STRUGGLE FOR ANSWERS

Lewin L. Williams

There has to be a shared responsibility between the church and the theological institutions in the search for answers to our Caribbean reality. While this Consultation has sought at this time to place some of the issues on the table, the search for answers takes both time and cooperation in effort. Perhaps what we are most able to do in this wrap-up session, is to bring more clarity to the issues raised over these past few days, and attempt to list some broad areas in which the diligent search for answers may begin.

REFLECTION AND ACTION AS THEOLOGICAL EDUCATION

There has to be an understanding that the Theological College is an academic institution. While it is true that because of the relationship between the Theological College and the University there may be an overemphasis on grades, even without the University, the Theological College demands its own academic excellence. It has no choice since, like the University, it is an academic institution.

On the other hand, the Theological College cannot afford the luxury of pure theory, because theory, according to Aristotle, tends to be knowledge for its own sake. It is here that "praxis", in its liberation sense, takes on particular importance in theological education. The Theological College must provide education for the transformation of society. Here, then, we develop our concern for the church's understanding of its own responsibility in the area of justice, and the relationship between justice and ministerial formation. For if the church's understanding of ministerial formation is training for the maintenance of some kind of "spiritual" status quo, the College and the church may be at odds regarding the definition of ministerial formation.

Many participants are asking, "What shall we do?" And perhaps asking that question is the easiest thing to do under the circumstances. Yet there are no easy and immediate answers. The struggle for answers is the struggle for self-understanding, and that takes time. Yet it is something that both the church and its training institutions must work at with some amount of diligence.

GENERAL OBSERVATIONS

Theological students tend to arrive at the College with an infantile faith system that must, of necessity, be dismantled for the building of a more mature one. That fact alone raises some questions:

- Does theological education today provide a support system against the possibility of a total loss of faith?

- On the other hand, what kind of responsibility does the church take for the kind of recruits who appear at the training institution? How are they prepared to be ready for the rigours of theological training?

Ministerial graduates seem unprepared for their working world, despite academic excellence in college. Could that condition owe its existence to some lack of emphasis on the importance of context? It is true that the Caribbean is only one small part of the world situation yet, even for that reason alone, contextual awareness is one of the most significant areas of theological education. Unless we maintain that which is truly Caribbean, we are a faceless, nameless nonentity on the world's stage.

- The individualistic trends that have become a part of the Caribbean reality are not themselves of origin. But even as rugged individualism is becoming normative for the Caribbean society, the Church does not realize how much it is impacted by the market place philosophy that has spawned such individualism. We need, then, to focus again on what is the true nature of the Caribbean reality by paying painstaking attention to our basic cultural heritage.

- The theological institution needs to take far more seriously, not only the development of a Caribbean theology, but also make certain that it is a consciousness that permeates the total curriculum for ministerial formation.

There is an observable retardation in the interest in ecumenism once the UTCWI experience is over. Is this due simply to the denominational labour demands upon the pastor's life, or is there a denominational perception of some inherent danger to faith and doctrine in ecumenical sharing? Whatever we deem the reasons to be, if Caribbean unity is crucial to Caribbean experience, the Church as the "soul" of the Caribbean community can hardly bypass ecumenism as a source for making leadership impact within the Caribbean. It is at this juncture that the work of the Caribbean Conference of Churches (CCC) proves to be of inestimable importance.

If our response to the *Missio Dei* is feeble, the reason may very well lie in the fact of our colonial past in which we see ourselves not as agents of mission, but always as objects of mission.

RECOMMENDATIONS

- Since the 1980s may very well rate as the "lost decade" for Caribbean theology, consultations such as this one may help to place the enterprise back on track. It is therefore recommended that another consultation be planned for 1995.

- Since some of the dormancy in the area of Caribbean theology may be due to lack of grassroots participation, we need to look to an "agency of integration" between theory and praxis. It is therefore recommended that an organization similar to the Costa Rican Department of Ecumenical Investigations (DEI) be developed in the

Caribbean, and that we seek assistance of funding agencies towards this venture.

- At present, not all theological colleges in the Caribbean are part of the Caribbean Association of Theological Schools (CATS). This situation exists because the university has not allowed some schools to function as one of its departments. It is recommended that we facilitate more broad-based dialogue among the Caribbean theological schools, and that CATS be extended beyond the servicing of the university's needs, to incorporate all the theological schools into its membership. Under this rubric, there should be a regular exchange among the teachers and students of these institutions, and Cuba should be particularly invited to participate.

- It is recommended that the theological institutions investigate the area of "education by extension" in which persons are not necessarily trained within the walls of the institution, but possibly on their own turf and work site.

- The church should be encouraged to develop centres that may serve as an ecumenical support system where dialogue is continuous and healthy.

- We should see the *Misio Dei* as our true focus because it is really God's mission that we are called to promote.

PRESENTATION TOPIC: 1

CARIBBEAN REALITY

Presenters: Noel Titus and Barry Chevannes

GROUP 1

Issues raised were
* *Sexuality*
* *The Common-Law Relationship*
  This was felt to be not the ideal, but, nevertheless, a reality.
* *Church's Ministry of Caring*
  Theology will have to develop a ministry of caring for those engaged in this pattern of relationship. Caribbean Theology will have to redefine marriage in the Caribbean context, taking into account its culture.
* *Limited Resources — Another Caribbean Reality*
  More interaction among theological institutions of the Caribbean region by student and faculty exchange. An annual retreat for the faculty from the theological institutions of the region was proposed.

GROUP 2

The group devoted all of the discussion time to the problem of recruiting persons for the full-time ministry. It was felt that there was need for serious discussion about this problem at all levels of the church. As part of looking at the problem, there was a feeling that if the church is not open to women for ministry, we may have, in time, to close down many theological colleges.

The group recognized that some new thinking on ministry will have to be considered, when, for example, both husband and wife are ordained and have churches of their own. When one has to move the other spouse has a problem, and limited work to move to as well.

It was also felt that there was need for a serious lay-training programme to ensure lay persons serve the church more effectively. This could be attempted on an ecumenical basis.

Why are more young people not offering themselves for the ministry as a full-time vocation? The group identified some of the reasons:

- Young people are not always challenged sufficiently for ministry — even to make sacrifices.
- Very often parents discourage their youngsters from considering the ministry as a vocation.
- Sometimes members of the clergy discourage young people.
- Young people observe how the church treats and cares for its ministers. If adequate caring is not recognized, they are turned off.
- Young people often become aware of the church's mismanagement and they are turned off by that.

There was a strong feeling that young people should be challenged to be followers of Christ — before any challenge is given to them about recruitment for ministry.

The discussions of the groups can be summarized in one sentence: *A methodology needs to be worked out for recruitment.*

PRESENTATION TOPIC: 2

METHOD IN

CARIBBEAN THEOLOGY

Presenter: Theresa Lowe-Ching

GROUP 1

Stimulating discussions ensued in Group 1 pertaining to Dr Theresa Lowe-Ching's presentation, "Method in Caribbean theology". Members of the group were generally impressed with the presentation and in particular, with the distinct effort made to link theory with practice, and with the response by Dr Lewin Williams. It is said that theology is the constant reflection of the people. The theology which has been spouted about provides the text which has to be applied to the reality or the context of the people.

How do we move towards integrating practice with theory and how do we teach in our seminaries?

In answering these questions it was suggested that a dialectical approach be used so as to stimulate the consciousness of the students. This has to be done at the expense of the old system where information was imparted without any input from the students. This kind of dialectical process also needs to be featured in our church where the members in the pew who do theology in their own right can participate in the Sunday morning sermonizing. One member said that in his congregation members have the right to participate in that way.

Following closely upon this is the fact that the "professional" theologians need to recognize the input of ordinary people. This is based on the fact that they also have experiences of God which shape their concepts of God.

So as to be relevant in our theologizing, we must use the needs of people as our point of entry into theology. People cannot be seen outside of their needs, whatever they may be. There is also a need for dialogue at all levels of theology. In developing this Caribbean theology, we must use the experiences and perceptions of all the peoples. This is achieved only through meaningful dialogue. The arts was suggested as a useful medium, since the Caribbean islands compete in this area.

The issue of a theology of imposition was also raised. The main point here being that some are afraid of developing a Caribbean theology because it seems to imply that they are no longer a part of the global system, i.e. by developing our theology we become somehow marginalized. Further, some people hold the view that to speak of God in their own terms is to stray from their own denominational status, which in some cases have come down through the ages. We need to affirm that each person's experience dictates the God she or he worships. Further, it is difficult to see God through the experience of other people.

A very important issue was raised as to the "voicelessness" of the Caribbean church about the Guyana situation. The same can also be said for Haiti. It is believed that the church refuses to address the oppressive forces which exist within the region. In a real sense it is a refusal to look at ourselves and address our deficiencies.

A concern was raised as to what system are we sending our well-formed ministers into. It has been observed that when these ministers go into the existing system, they cannot operate as they have been trained. Therefore they fall back into the existing molds and perpetuate the old system. In a very real sense a Caribbean theology must liberate us from colonial structures such as these.

At the end of our session the question was asked as to whether we need a Caribbean theology or Caribbean theologies. Further, do we want to call this a Caribbean theology since to call it Caribbean might relegate it to a secondary position.

GROUP 2

In this group there were more questions than answers, questions such as:

- How do we go about doing Caribbean theology?

The major response to this was pointed out to be by listening: to the people whom we serve, minister to, teach, or have some influence on.

- Do our structures allow for listening?
- Does this Consultation have enough time for listening and speaking back?
- Is a period of fifteen out of ninety minutes adequate for real listening?
- Is the structure of this Theological Consultation a contradiction of we are saying? For instance, what about representation of popular groups such as Rastafarians and Pentecostals?
- Do we listen enough to recognize the theological positions of the people when they express them?
- Is there a need for clergy to take seriously the desire, hopes, fears and aspirations of our people when they articulate them?
- Is there need to examine how our sermons, services, prayers, Bible studies, etc., are shaped and planned? Are they in response to needs expressed, or are they what the clergy leaders think is good for them?
- Does the leader, being in charge at all times, inhibit the spontaneous articulation of theology? Should the priest or pastor not take a back seat sometimes, thereby facilitating the process of articulation? Do Caribbean lay people feel comfortable to challenge the clergy face to face? (The general response to this is in the negative, though it was felt that a lot of challenging goes on outside the hearing of the clergy.)
- Do the traditional styles of ministry inhibit the expression of theology?
- Should theological training not involve students going out, maybe for a year, "to listen" and "to firm up" their theology and then return to the seminary to continue their studies?
- Are theological colleges too grade oriented? Are students under too much pressure?

It was felt that there is a need for more flexibility in theological programmes, and more time for reflection in seminaries. The group, however, recognized some problems in accommodating adequate reflection time at seminaries:

- Lecturers have syllabuses to complete.
- Churches make demands of seminaries.
- Churches have expectations of students
- Our alignment to the university places more pressure upon students. The university has a high academic emphasis and the university programme and the seminary programme run concurrently.
- Students want to graduate with a degree or they feel inferior to those who have degrees. Their preoccupation with the academics provide limited time for listening.

It was also felt that our present religious leaders could be of greater help to young clergy and lay people by listening some more, thereby facilitating expression, growth, building of confidence and so on.

Since funds are available from the World Council of Churches for the publication of works, seminaries should make use of these opportunities. They could play a coordinating role and encourage Caribbean persons with special gifts to write and to publish their works for sharing within the region, so that more people can benefit from what they have to say.

GROUP 3

The group decided to comment generally on points that had been raised both during the opening session and during the first morning session.

On Caribbean theology as essentially liberation theology: a note of caution was expressed. We must be careful not to simply copy the liberation theologies of Latin America or elsewhere. Caribbean theology need not bear the label "liberation theology" to be liberating. Sources of Caribbean theology must include not only those mentioned by the speakers (Bible etc.) but also important Caribbean figures who do not belong to the "religious" sphere, e.g., Sparrow, Naipaul, Marley, Garvey, Walcott, Rodney, Granger and others, including women (no women were actually named). These persons articulate Caribbean experiences in various ways.

On contextualization of Caribbean theology several points were made:

- The local congregation should be the major focus of contextualization: true theological school participation.
- Questions arose as to the difficulties of bridging the gap between the Theological College and the local congregation.
- A major obstacle to contextualization is the failure of theologians to listen to what is taking place in the local congregations and the theologies being elaborated there.
- The shift from education of ministers to education of ministries (Dr Ham) is important for contextualization and will help to move theology beyond the walls of the Theological College.
- Incorporation of the experience of those Caribbean persons who have Asian roots is crucial for contextualization.
- The question of a possible contribution from representatives of Caribbean religious faiths was also raised.
- Finally, several persons commented on the UTCWI in relation to the task of contextualization. There was need expressed for greater clarity about what students are being prepared for, and recognition of the fact that college education often has the effect of removing students from their social and cultural roots so that when they return to their congregations they are no longer trusted by the people. It was acknowledged, however, that the College ought not be made to accept responsibility for producing particular kinds of leaders if the churches themselves do not set the tone in this matter. One suggestion was that churches need to shift from a maintenance mode to a missionary mode of existence so that people may learn the task of mission.

PRESENTATION TOPIC: 3

THE GLOBALIZATION OF

THEOLOGICAL EDUCATION

Presenter: Winston D. Persaud

GROUP 1

Our theology begins in our interpretation of the space between ourselves and God. Who controls that space determines, to a large extent, who controls the interpretation, and who interprets or defines the theology.

What does the Caribbean have to offer to global theological education?

While the group generally appreciated a lot of what Dr Persaud said, some people felt a bit of discomfort with the North American model which he was using. It was felt that the Caribbean had a uniqueness which must be affirmed, but that there was need for a meeting point — some kind of commonality among all religions and cultures that could be called Caribbean. It was pointed out that UTCWI does not educate ecumenically, but uses a common ground approach.

It was also felt that it is presumptuous to talk about the globalization of theology when, as a Caribbean people, we have not yet clarified our own theological position. We need to work that out before we talk about globalization.

Two questions were raised:

- Do we have to wait for what happens outside the Caribbean, before we respond?
- What about Caribbean initiative?

It was felt that denominational structures often serve as stumbling blocks to initiative.

Another question asked was:

- Do we as different denominations, ever discuss doctrines, or do we avoid it?

It was pointed out that such discussions take place at the seminaries.

It was recognized that there are many things denominations have in common, e.g., churches with great doctrinal differences have united and marched together for a cause with a common ground and on a matter like justice. The problem was that they do not usually keep in regular contact.

Strong feelings were expressed about the difficulty experienced in finding a formalized model of globalization. It was felt, however, that all doors must be kept open.

There is no easily definable Caribbean context

The group recognized a difficulty in defining a Caribbean person, a Caribbean identity. There was also a real difficulty in defining a Caribbean theology. Whatever Caribbean theology is, it has to deal with that difficulty.

GROUP 2

What does Caribbean theology have to offer, and what should it offer to global theology?

The group began by considering the first question that Dr Persaud had proposed for discussion. As the discussion began, it became apparent that several participants were unclear about what precisely is meant by "globalization", and so we set about attempting to clarify that notion.

Some concrete examples of attempts at globalization helped to throw some light on the question. It was pointed out that globalization must be carefully distinguished from "unipolarization". The former can be considered in a position way, as a basic openness can enrich one's own understanding, and can aid the process of finding some common ground among Christians of different backgrounds, so that a common Christian response or critique can be given to various issues in the global society.

Difficulties in articulating what Caribbean theology has to offer to global theology

The opinion was expressed that globalization of theological education presupposed a model of such education that, based on that of the University, is not only inappropriate, but unaffordable in the Jamaican and Caribbean context. Churches may be forced to choose some other model of theological education.

The example of UTCWI was raised, and the opinion was expressed that not sufficient emphasis was placed in the curriculum on the specific contributions of Caribbean theologians.

The migration of Caribbean theologians is also a difficulty. There is need to get some of them back, though it was also pointed out that this might be a mixed blessing, especially if such persons have become alienated from the Caribbean.

The contribution of Caribbean theology to global theology

The importance of the uniqueness of our own historical experience was stressed. However, this needs to be appropriated in all its aspects if persons are to have the self-confidence to make a contribution that is born of such experience.

BIBLIOGRAPHY

Amirtham, Samuel, and Robin Pryor, eds. 1990. *Resources for Spiritual Formation in Theological Education.* Geneva: World Council of Churches Programme on Theological Education.

Battle, Rosario. 1989. "A structure that encloses formation for mission". In *Ministerial Formations, Mission, Today's World,* edited by John S. Pobee. A collection of papers from a Consultation on Ministerial Formation for Mission Today, Limuru, Kenya.

Baum, Gregory. 1981. "Ecumenical theology: a new approach". *The Ecumenist* 19, no. 5 (July-August).

Browning, Don S. 1986. "Globalization and the task of theological education in North America". *Theological Education* 23, no. 1 (Autumn).

Caribbean Policy Development Centre. 1992. *Challenges in Caribbean Development.* Barbados: CPDC

CARICOM. 1991. "Guidelines for economic development: strategy for CARICOM countries into the 21st century". CARICOM Regional Economic Conference Document.

Carter, Martin. 1966. "Looking at your hands". In *Caribbean Voices: An Anthology of Caribbean Poetry,* Vol. 1. Selected by John Figueroa. London: Evans Brothers Ltd.

Cuthbert, Marlene, ed. 1971. *Role of Women in Caribbean Development: Report on Ecumenical Consultations.* Barbados: Christian Action for Development in the Eastern Caribbean (CADEC) Publications.

Davis, Kortright, ed. 1977. *Moving into Freedom.* Barbados: Cedar Press.

Davis, Kortright. 1982. *Mission for Change: Caribbean Development as Theological Enterprise*. Bern: Verlag Peter Lang.

Davis, Kortright. 1990. *Emancipation Still Comin': Explorations in Caribbean Theology*. New York: Orbis Books.

D'Costa, Gavin. 1988. "Against religious pluralism". In *Different Gospels*, edited by Andrew Walker. London: Hodder & Stoughton.

de Santa Ana, Julio. 1991. "Theses on theological education". Paper delivered at the Conference on Theological Education in Situations of Bare Survival, 14-18 July, Managua, Nicaragua, translated by Kield Renato Ling.

Erskine, Leo. 1981. *Decolonizing Theology: A Caribbean Perspective*. New York: Orbis Books.

Felder, Cain Hope. 1990. *Troubling Biblical Waters: Race, Class and Family*. New York: Orbis Books.

Freire, Paulo. 1972. *Pedagogy of the Oppressed*. Harmondsworth: Penguin.

Gayle, Clement. 1991. "History of the United Theological College of the West Indies". Twenty-fifth Anniversary Souvenir Magazine of the United Theological College of the West Indies.

Hamid, Idris. 1973. *In Search of New Perspectives*. Bridgetown, Barbados: Conference of Churches Publication.

Hamid, Idris, ed. 1973. *Troubling of the Waters*. San Fernando, Trinidad: Rahaman Printer Ltd.

Hamid, Idris. 1977. *Out of the Depths*. San Fernando, Trinidad: St Andrews Theological College.

Heim, S. Mark. 1990. "Mapping globalization for theological education". *Theological Education* 26, Supplement 1 (Spring).

Heniff, Nesha. 1988. *Blaze a Fire: Significant Contributions of Caribbean Women*. Toronto: Sistren Vision.

Hinkelammert, Franz. 1992. "Abya Yala and the current world situation". Paper delivered at the Consultation on Theological Education in Abya Yala, 20-24 July, San José, Costa Rica.

Hromadka, Joseph. 1990. "The witness of the church in changing Czechoslovakia". Lecture, 9 October, Princeton Theological Seminary.

Knight, Franklin. 1978. *The Caribbean*. New York: Oxford University Press.

Krieger, David J. 1991. *The New Universalism: Foundations for a Global Theology*. New York: Orbis Books.

Lamb, Matthew. 1976. "The theory-praxis relationship in contemporary Christian theologies". *Catholic Theological Society of America Proceedings* 31.

Leo-Rhynie, Elsa. 1992. "Women and development studies: moving from the periphery?" Lecture presented at the Women and Development

Studies Tenth Anniversary Symposium, 8-10 December, University of the West Indies, Mona.

McAfee, Kathy. 1991. *Storm Signals: Structural Adjustment and Development Alternatives in the Caribbean.* London: Zed Books.

Naipaul, Shiva. 1988. *An Unfinished Journey.* Bergenfield, N.J.: Viking Penguin.

Neil, Stephen. 1964. *A History of Christian Missions.* Harmondsworth: Penguin Books.

O'Brien, John. 1992. *Theology and the Option for the Poor.* Minnesota: The Liturgical Press.

Oden, Thomas. 1987. *Becoming A Minister.* New York: Crossroad.

Owens, Joseph. 1973. "Rastafarians of Jamaica". In *Troubling of the Waters,* edited by Idris Hamid, pp 165-70. San Fernando, Trinidad: Rahaman Printer Ltd.

Panton, Vivian. 1992. *The Church and Common-law Union: A New Response.* Kingston, Jamaica: By the author.

Persaud, Winston D. 1991. *The Theology of the Cross and Marx's Anthropology: A View from the Caribbean.* New York: Peter Lang.

Pobee, John, ed. 1989. *Ministerial Formation, Mission, Today's World.* A collection of papers from a Consultation on Ministerial Formation for Mission Today, Limuru, Kenya.

Roberts, George. 1955. "Some aspects of mating and fertility in the West Indies", *Population Studies* 8, no. 3.

Ruether, Rosemary Radford. 1970. "Women's liberation and theological perspective". In *Women's Liberation and the Church,* edited by Sarah Bartly Doely. New York: Association Press.

Ruether, Rosemary Radford. 1983. *Sexism and God Talk: Toward a Feminist Theology.* Boston: Beacon Press.

Russell, Horace. 1991. "A brief account of the formation of the United Theological College of the West Indies and its development". Mimeo.

Schneiders, Sandra. 1991. *Beyond Patching.* New York: Paulist Press.

Segundo, Juan Luís. 1976. *The Liberation of Theology.* New York: Orbis Books.

Segundo, Juan Luís. 1984. *Faith and Ideology.* New York: Orbis Books.

Severino Croatto, J. 1990. "Spiritual formation and critical study". In *Resources for Spiritual Formation in Theological Education,* edited by Samuel Amirtham and Robin Pryor. Geneva: World Council of Churches Programme on Theological Education.

Smith, Ashley. 1973. "The religious significance of Black Power in Caribbean churches". In *Troubling of the Waters,* edited by Idris Hamid, pp 83-104. San Fernando, Trinidad: Rahaman Printer Ltd.

Smith, Ashley. 1984. *Real Roots and Potted Plants: Reflections on the Caribbean Church*. Jamaica: Eureka Press.

Taylor, Mark. 1989. *Erring: A Post-modern A/Theology*. Chicago: University of Chicago Press.

Watty, William. 1981. *From Shore to Shore: Soundings in Caribbean Theology*. Barbados: Cedar Press.

Weir, J. Emmette. 1991. "Towards a Caribbean liberation theology". *Caribbean Journal of Religious Studies* 12, no. 1 (April).

West Indian Commission. 1992. *Time for Action — Report of the West Indian Commission*. Black Rock, Barbados: The West Indian Commission.

Williams, Daniel Day. 1961. *The Minister and the Care of Souls*. New York: Harper and Row.

Williams, Eric. 1970. *From Columbus to Castro: The History of the Caribbean 1492-1969*. London: André Deutsch.

Williams, Lewin. 1991. "What, why and wherefore of Caribbean theology", *Caribbean Journal of Religious Studies* 12, no. 1 (April).